MICHAEL J. H. TAYLOR
and JOHN W. R. TAYLOR

Missiles of the World

Charles Scribner's Sons
New York

Copyright © 1972 Ian Allan

All rights reserved. No part of this book
may be reproduced in any form without the
permission of Charles Scribner's Sons.

A—12.72 (I)

Printed in Great Britain
Library of Con...

SBN 684–131...

358.17	Taylor, Michael John
TAY	Missiles of the
1972	world [by] Michael J. H. Taylor and John W. R. Taylor.

Foreword

This companion volume to *Military Aircraft of the World* and *Civil Aircraft of the World* contains details and photographs of all guided missiles known to be in service or under development throughout the world. Even the medium-range and intercontinental ballistic missiles being produced by China are included, together with the full range of guided weapons deployed in thousands by Russia and its allies. Some of the photographs have never before been published. All of the facts and figures are as up-to-the-minute and accurate as possible, having been compiled with the help of correspondents, manufacturers and official organisations in more than a dozen countries.

Here, for the engineer and military student, are standardised descriptions of weapons like 'Styx', which sank the Israeli destroyer *Eilat*; Swingfire, which can be launched round corners to knock out the heaviest battle tanks; the mighty Russian SS-9, which can deliver a 25-megaton warhead, a pattern of smaller nuclear charges or a 'space bomb'; and Britain's Seacat, which is so effective that no ship of the fifteen navies that deploy it has ever been attacked. In every case, we have tried to present the information in a way that can be understood by non-technical readers, who cannot fail to be awed by weapons so terrifying that they have made a major war impossible.

MJHT
JWRT

Soviet intercontinental ballistic missile in its underground silo launcher

AAM-1

Air-to-air missile. In production and service.

Prime Contractor: Mitsubishi Jukogyo Kabushiki Kaisha.
Guidance and Control: Infra-red homing guidance system.
Warhead: High-explosive.

Development and Service
Mitsubishi Heavy Industries developed this

Japan

air-to-air missile specifically to replace the Sidewinder on F-86F and F-104J interceptors of the Japan Air Self-Defence Force. Deliveries began in November 1970, and half of the currently-planned total of 330 missiles were expected to be in service by the beginning of 1972. No details of the AAM-1 have been released for publication, except that it is a pursuit-course weapon, limited to attack from the rear of the target.

AAM-2

Air-to-air missile. Under development.

Prime contractor: Mitsubishi Jukogyo Kabushiki Kaisha.
Guidance and Control: Nihon Electric Company infra-red homing guidance system.
Warhead: High-explosive.

Japan

Development and Service
The AAM-2 is being developed by Mitsubishi as a replacement for the AAM-1 from about 1973. No details are available, except that it will overcome the limitations of the AAM-1 by its ability to home on the target from any direction.

ACRA

Semi-automatic anti-tank missile. Under development.

Prime contractor: Direction Technique des Armements Terrestres.
Airframe: Cylindrical body with cruciform flip-out tail control surfaces of narrow chord and four tiny anti-roll fins. Conical nose with rounded tip.
Guidance and Control: Infra-red beam-riding guidance system. Control by tail surfaces.
Warhead: Hollow-charge.
Length: 3ft 11¼in (1.20m).
Body diameter: 5.59in (142mm).
Launch weight: 77lb (35kg).
Cruising speed: 1,118mph (1,800km/h).
Max range: 9,840ft (3,000m).

Development and Service
The semi-automatic ACRA (Anti-Char RApide) is launched from a smooth-bore gun of 142mm calibre and carries a warhead powerful enough to destroy any known tank with a single hit. It was developed by the Atelier de Construction de Puteaux (APX), which is part of the industrial complex of the DTAT, together with a shorter, unguided anti-personnel weapon which can be fired from the same gun. ACRA leaves the muzzle of the gun, mounted on a combat vehicle, with an initial velocity of 492ft (150m)/sec. This is increased to 1,640ft (500m)/sec by the missile's own motor, with the result that it covers its maximum range in under seven seconds. The possibility of the target taking counter-action against the ACRA launch vehicle is thus reduced to a minimum.
 ACRA travels along an infra-red director

France

beam emitted by a laser. This beam is slaved to the operator's optical sight, which need only be held on target, with no manual steering requirement. The missile corrects its course automatically, by sensing deviations from course by comparison with the axis of the director beam—a form of guidance which is immune to jamming and offers a high degree of accuracy. Testing is scheduled for completion during 1973, ten years after development of the weapon was initiated. ACRA will then be deployed initially on AMX-10M armoured vehicles of the French Army.

Advanced Terrier (RIM-2F) USA

Ship-based surface-to-air missile. In service

Prime contractor: General Dynamics Corporation.
Powered by: Allegany Ballistics Laboratory tandem two-stage solid-propellant rocket motors.
Airframe: Tandem two-stage design. Cylindrical missile body, with long ogival nose, cruciform long-chord wings of constant narrow span with cruciform tail control surfaces immediately to their rear. Cylindrical tandem booster, of larger diameter, with cruciform cropped-delta fins at rear, indexed in line with missile surfaces.
Guidance and Control: General Dynamics (Pomona) beam-riding guidance with semi-active terminal homing. Control by movable tail surfaces.
Warhead: Naval Ordnance Laboratory high-explosive warhead, with proximity fuse.
Length: 27ft 0in (8.23m).
Body diameter: Missile 1ft 0in (30cm) Booster 1ft 4in (40.6cm).
Wing span: 1ft 8in (0.51m).
Launch weight: 3,000lb (1,360kg).
Max range: over 20 miles (32km).

Development and Service

The original Terrier missile (RIM-2A, formerly SAM-N-7) was evolved from an experimental vehicle named Lark and became operational on ships of the US Navy in 1956. It was followed seven years later by Advanced Terrier, with new beam-riding guidance which offered greatly increased effectiveness against low-flying aircraft and multiple targets, and also gave the missile a surface-to-surface capability. Externally, the Advanced weapon differs from Terrier mainly in having

strake-like wings instead of the former small cropped-delta surfaces.

Advanced Terrier continues to arm many US attack carriers, cruisers and frigates, as well as one Dutch and four Italian cruisers. It is being replaced by Standard Missile.

Aegis
USA

Surface-to-air ship-based missile. Under development.

Prime contractor: Radio Corporation of America.
Powered by: Dual-thrust solid-propellant rocket motor.
Airframe: Cylindrical body, with long-chord cruciform wings, indexed in line with cruciform tail control surfaces. Ogival nose-cone carrying four small sweptforward surfaces, each with tip-mounted antenna.
Guidance and Control: Semi-active radar guidance. Control by tail surfaces.
Warhead: High-explosive.

Development and Service

Aegis was known originally as the advanced surface missile system (ASMS) and will be the US Navy's primary anti-aircraft defence system from about 1976 into the '80s. It is being planned as an area-defence system, able to protect an entire task force. Initial deployment will be on board a new class of destroyers and frigates, with possible further installation on *Nimitz* class aircraft carriers. System features will include an electronic scanning radar able to look almost instantaneously in all directions, a dual-purpose launcher able to carry either Aegis missiles or Asroc (see page 15), the new UYK-7 naval tactical data system computer, and microwave radars for target illumination.

Development of Aegis has been underway since 1964. An engineering development contract was awarded to RCA in December 1969 and flight testing of the system is expected to begin in 1973. Bendix is contractor for missile design and support equipment; Raytheon is responsible for the microwave target illuminator and other electronics. The Aegis missile is expected to be similar in size to the RIM-66A medium-range Standard Missile (see page 149), and the basic Aegis system may use RIM-66s until the definitive missiles are available for service.

Agile

USA

Short-range air-to-air missile. Under development.

Guidance and Control: Infra-red guidance system.
Minimum range: 3,000ft (915m).

Development and Service
Agile is the 'dogfight' missile which the US Navy plans to develop as armament for the Grumman-F-14 Tomcat, following experience with earlier types of air-to-air missile in Vietnam. It will be highly-manoeuvrable, to deal with fast-turning enemy aircraft at close range, and may also be produced for the USAF following cancellation of the latter's AIM-82A 'dogfight' missile for the McDonnell Douglas F-15. The Naval Weapons Center, China Lake, California, is responsible for the Agile project. A contract for the infra-red seeker has been awarded to Hughes Aircraft Company.

Albatros

France/Italy

Air-to-surface anti-shipping missile. Under development.

Prime contractors: SA Engins Matra (France) and Oto Melara SpA (Italy).
Powered by: Turboméca Arbizon III turbojet, rated at 882lb (400kg) st.
Airframe: Cylindrical body with pointed nose-cone. Four semi-circular engine air-intake ducts, each supporting a cropped-delta wing, equi-spaced around body. Cruciform tail control surfaces indexed in line with wings.
Guidance and Control: Thomson-CSF terminal guidance. Control via tail surfaces.
Warhead: High-explosive.
Max range: 37-50 miles (60-80km).

Development and Service
This stand-off weapon can be launched from aircraft such as the Atlantic and Nimrod maritime reconnaissance-bombers from well outside the range of most current and projected ship-to-air defence systems. Basically, it combines the rear portion of the Otomat ship-to-ship missile (page 87) with the forward section of Martel (page 76); but, being air-launched, it does not need Otomat's boosters. After launch, it descends to a very low altitude, which is maintained by a radio-altimeter. Then, like Otomat, it climbs before making a terminal dive into the target. Albatros is expected to be ready for service by 1973 and will be suitable as armament for large anti-submarine helicopters in the class of the French Super Frelon.

Mock-up of Albatros

Albatros
Italy

Surface-to-air missile system. Under development.

Prime contractor: Selenia SpA.

Development and Service
Not to be confused with the Albatros air-to-surface anti-shipping missile, Selenia's Albatros weapon system is intended to provide naval craft with all-weather point defence against attack by aircraft and missiles, down to very low altitudes. It consists of either a Ferranti GA-10 digital gun fire control system or Elsag NA-10 gun fire control system, Selenia Orion RTN 10 tracking radar, a missile launching system and one or two anti-aircraft guns, and Sparrow (AIM-7) missiles (see page 138). The missiles are modified by the use of folding wings and clipped tail-fins, to fit into the launcher, and by the addition of a rapid run-up capability. Only a single operator is required for the Albatros system, to carry out radar control functions, missile launch and gun fire control. Defensive cover is provided from the maximum range of the missiles (over 6 miles; 10km) down to the minimum effective range of the guns (about 1,000ft; 300m).

Selenia received an official contract to begin development of the Albatros system in early 1968. Successful firing trials against target drones in Sardinia are being followed by operational evaluation on board a ship of the Italian Navy. Production was expected to start in 1972.

Alkali
USSR

Air-to-air missile. In service.

Powered by: Solid-propellant rocket motor.
Airframe: Basically-cylindrical body, with large rear-mounted cruciform wings of delta shape, indexed in line with small cruciform foreplanes. Control surfaces in wing trailing-edges.
Guidance and Control: Radar homing system. Control via wing trailing-edge surfaces.

Development and Service
'Alkali' continues to be carried by all-weather interceptor versions of the MiG-19, in service with some of Russia's allies and friends. It was the first-generation missile armament of this fighter and of the original version of the Sukhoi Su-9.

Alkali missiles under the wings of a MiG-19

Anab missiles under the wings of an Su-11

Anab   USSR

Air-to-air missile. In service.

Airframe: Cylindrical body, fitted with large cruciform tail-fins and small canard cruciform foreplanes.
Guidance and Control: Alternative infra-red and semi-active radar homing guidance systems. Control by movable foreplanes.
Warhead: High-explosive.
Length: IR version 13ft 5in (4.1m).
Radar homing version 13ft 1in (4.0m).

Development and Service

Like the USAF, the Soviet Air Force has learned by experience the value of carrying both infra-red homing and radar homing missiles on its interceptors, for use under varying weather conditions and in different tactical situations. The Sukhoi Su-9 and Yakovlev Yak-28 all-weather fighters of the Protivo-Vozdushniya Oborona Strany (National Anti-Air Defence force) are thus armed usually with one 'Anab' of each type on underwing racks. So is the new single-seat twin-jet Su-11.

Two AS.11 missiles mounted on a Scout helicopter

AS.11. (AGM-22A)

France

Air-to-surface tactical missile. In production and service.

Prime contractor: Aérospatiale, Division Engins Tactiques.
Powered by: Two-stage solid-propellant rocket motor.
Airframe: Cylindrical body, with cruciform sweptback wings and rounded nose.
Guidance and Control: Wire guidance. Control by varying sustainer thrust through two side nozzles.
Warhead: High-explosive, high-fragmentation anti-personnel or armour-piercing, of several types.
Length: 3ft 11in (1.20m).
Body diameter: 6½in (0.16m).
Wing span: 1ft 7½in (0.50m).
Launch weight: 66lb (29.9kg).
Average cruising speed: 360mph (580km/h).
Range limits: 1,650-9,840ft (500-3,000m).

Development and Service

This is the air-to-surface version of the SS.11 tactical missile, to which it is generally similar (see page 145). When carried by helicopters, it is normally controlled by means of a specially-developed stabilised sight. The AS.11 is in widespread service with the French armed forces and the air forces of other nations, including the UK and USA, by which it is designated AGM-22A.

AS.12 on the underwing launcher of a French Navy Alizé

AS.12
France

Air-to-surface missile. In service.

Prime contractor: Nord-Aviation/Aérospatiale, Division Engins Tactiques.
Powered by: Two-stage solid-propellant rocket motor.
Airframe: Cylindrical body, with cruciform wings and bulged ogival nose.
Guidance and Control: Wire-guidance. The missile is spin-stabilised and steered by varying the sustainer thrust through two side-nozzles. The AS 12 can be adapted to use TCA guidance, as described for the AS.30 (page 00).
Warhead: High-explosive, weighing 66lb (30kg).
Length: 6ft 1.9in (1.87m).
Body diameter: 7in (0.18m).
Wing span: 2ft 1½in (0.65m).
Launch weight: 167lb (75kg).
Speed at impact: 210mph (335km/h) when launched at 230mph (370km/h).
Max range (in relation to aircraft): 18,000ft (5,500m).

Development and Service
By scaling up the basic SS.11 and AS.11 to produce the SS.12 and AS.12, Nord-Aviation was able to fit a much larger warhead and so create a missile that would be effective against fortified positions as well as tanks, ships and other vehicles. The armour-piercing version will penetrate more than 1.5in (40mm) of steel armour and explode on the other side. Alternatives include an anti-tank shaped charge and a pre-fragmented anti-personnel type. Most missiles in service use manual steering, although TCA is available. They are carried by Alizé and Atlantic anti-submarine aircraft and are alternatives to Albatros on the underwing racks of the Nimrod. Four AS.12s can be carried by the Anglo-French Lynx helicopter, together with an optical stabilised sight.

AS.30 (*left*) and AS.20 (*right*) under the port wing of a Vautour all-weather attack aircraft of the French Air Force

AS.20
France

Air-to-surface missile. In service.

Prime contractor: Nord-Aviation/Aérospatiale, Division Engins Tactiques.
Powered by: Dual-thrust solid-propellant motor, Booster stage exhausts through two lateral nozzles, sustainer through nozzle at tail.
Airframe: Cylindrical body, with pointed ogival nose and cruciform wings which are canted to spin-stabilise the missile.
Guidance and Control: Radio command by pilot of launch aircraft, with aid of tracking flares at rear of missile. Control by jet-deflectors in sustainer nozzle.
Warhead: High-explosive, weighing 66lb (30kg), with contact fuse.
Length: 8ft 6½in (2.60m).
Max body diameter: 9¾in (0.25m).
Wing span: 2ft 7½in (0.80m).
Launch weight: 315lb (143kg).
Max speed: Mach 1.7.
Max range: 4.35 miles (7km).

Development and Service
Any aircraft capable of launching the AS.20 at a speed of Mach 0.7 or higher can be equipped with it, and more than 7,000 were produced by Nord-Aviation (now Aérospatiale) for the French Air Force and Navy and for four other countries. It is a standard air-to-surface weapon on the Mirage III, and the West German and Italian Air Forces mount AS.20s on their Fiat G91s. By means of a special adaptor, it can be fired as a training missile from aircraft equipped to carry the more powerful AS.30. The AS.20 can also be adapted to utilise the TCA optical aiming/infra-red guidance system described under the AS.30 entry.

AS.30

France

Tactical air-to-surface missile. In service.

Prime contractor: Nord-Aviation/Aérospatiale, Division Engins Tactiques.
Powered by: Two-stage solid-propellant rocket motor.
Airframe: Cylindrical body with pointed ogival nose. Canted cruciform sweptback wings, indexed in line with cruciform flip-out tail-fins.
Guidance and Control: Radio command, with optional TCA automatic guidance.
Warhead: High-explosive, weighing 510lb (230kg), with delay or non-delay fuse.
Length: 12ft 7in (3.84m) or 12ft 9in (3.89m) depending on warhead.
Body diameter: 1ft 1½in (0.34m).
Wing span: 3ft 3½in (1.00m).
Launch weight: 1,146lb (520kg).
Speed at impact: 1,005–1,120mph (1,620–1,800km/h).
Average range: 6.8–7.5 miles (11–12km).

Development and Service
Scaled up from the AS.20, with added flip-out tail-fins, this formidable missile is used by the French Air Force and Navy, the RAF, and the Air Forces of Germany, Israel, South Africa and Switzerland. It was designed to guarantee impact within a 33ft (10m) circle drawn around the aiming point when launched at least 6.2 miles (10km) from the target, without any need for the launch aircraft to approach within 1.8 miles (3km) of the target. This specification was exceeded by the production missiles, of which about 5,500 were delivered. They can be launched at any speed above Mach 0.45, with no upper limit.

Many of the AS.30s in service with the French Air Force have been adapted to utilise TCA (Télécommande Automatique) guidance, which frees the operator of having to 'acquire' the missile after launch and align it manually on the target by means of a small joystick control. Instead, the operator simply keeps an optical sight aligned on target. After launch, infra-red radiation from a flare on the missile is picked up by a goniometer, the reference axis of which is parallel to the axis of the optical sight. The goniometer passes to a computer information on any deviation of the missile from this axis; signals are then passed to the missile automatically over the guidance wires, to bring it parallel to the line of sight and keep it there. Tests have shown that TCA can keep a missile within less than 3ft 3½in (1.0m) of the direct line to the target.

AS.30L

France

Air-to-surface missile. Available for service.

Prime contractor: Aérospatiale, Division Engins Tactiques.
Powered by: Two-stage solid-propellant rocket motor.
Airframe, Guidance and Control: Similar to AS.30, but wings reduced in span and moved further aft.
Warhead: High-explosive, weighing 253lb (115kg).
Length: 11ft 9½in (3.60m).
Max body diameter: 1ft 1½in (0.34m).
Wing span: 2ft 11¼in (0.90m).
Launch weight: 838lb (380kg).

Development and Service
This lightweight version of the AS.30 was developed to equip aircraft in the class of the Fiat G91 fighter.

Ash missiles under the wings of an early version of the Tu-28P. Current operational Tu-28Ps each carry four missiles

Ash

USSR

Air-to-air missile. In service.

Airframe: Basically-cylindrical body. Large cruciform delta wings indexed in line with cruciform tail surfaces.
Guidance and Control: Alternative infra-red and semi-active or active radar homing guidance systems.
Warhead: High-explosive.
Length: IR version 18ft 0in (5.5m)
Radar-homing version 17ft 0in (5.2m).

Development and Service
Largest Soviet air-to-air missile in current service, 'Ash' is normally carried as a 'mix' of infra-red and radar homing versions by all-weather fighters of the Soviet Air Force. The big Tupolev Tu-28P twin-jet interceptor has underwing attachments for two missiles of each type.

Asroc

Asroc (RUR-5A)

Anti-submarine rocket. In service.

Prime contractor: Honeywell Inc.
Powered by: Naval Propellant Plant solid-propellant rocket motor.
Airframe: Cylindrical aluminium body, made in two halves and joining torpedo at front to motor at rear. Cruciform tail-fins.
Guidance and Control: Missile follows unguided ballistic trajectory, after launch towards target position predicted by shipboard sonar. Acoustic-homing torpedo warhead, or unguided nuclear depth charge.
Warhead: Alternative warheads include the General Electric Mk 44 Model 0 high-speed acoustic-homing torpedo, Aerojet-General Mk 46 Model 0 advanced acoustic-homing torpedo, Honeywell Mk 46 Model 1, or a nuclear depth charge developed by the Naval Weapons Center and Honeywell.
Length: 15ft 0in (4.57m).
Body diameter: 1ft 0in (0.30m).
Fin span: 2ft 6in (0.76m).
Launch weight: 1,000lb (450kg).
Range: 1-6 miles (1.6-9km).

Development and Service

Asroc development began in June 1956, and production was started three years later after successful firing trials. The weapon has been operational on destroyers, escort ships and cruisers of the US Navy since the Summer of 1961 and also arms the Japanese destroyer *Amatsukaze*.

In action, the target submarine's course, range and speed are worked out by shipboard computer within seconds of its detection by sonar. The eight-round standard launcher, or two-round modified Terrier missile launcher is aligned into firing position. The ship's commander then selects the missile with the most appropriate warhead and fires it. En route to the target, the Asroc sheds its rocket motor at a predetermined signal. Later, a steel band holding the airframe together is severed by a small explosive charge. The airframe then falls away, allowing the depth charge to drop into the water or the torpedo to be lowered to the surface by parachute.

USA

Asroc on a modified Terrier twin-launcher

Standard eight-round swivelling launcher for Asroc

Atoll under the starboard wing of a MiG-21PF

Atoll USSR

Air-to-air missile. In service.

Powered by: Solid-propellant rocket motor.
Airframe: Cylindrical body. Cruciform triangular control surfaces near nose, indexed in line with fixed cruciform tail-fins.
Guidance and Control: Infra-red homing guidance. Control by cruciform foreplanes and small gyroscopically-controlled tab at tip of trailing-edge of each tail-fin.
Warhead: High-explosive.
Length: 9ft 2in (2.80m).
Body diameter: 4.72in (12cm).
Fin span: 1ft 8¾in (0.53m).

Development and Service
Standard armament on Soviet Air Force and export versions of the MiG-21, 'Atoll' is similar to the American Sidewinder (page 133) in configuration, size and guidance system. It can be assumed that it is equally like the US weapon in performance and operational limitations. A number of other aircraft have been seen carrying 'Atolls' on underwing mountings, including the Yak-28P.

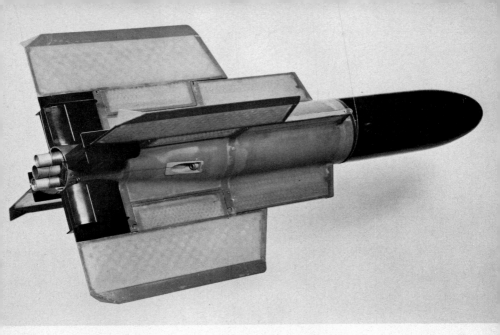

Bantam   Sweden

Light anti-tank missile. In production and service.

Prime contractor: Aktiebolaget Bofors.
Powered by: Two-stage solid-propellant rocket motor.
Airframe: Cylindrical body, with rounded ogival nose and rear-mounted cruciform wings, made largely of glassfibre-reinforced plastics. Vibrating spoilers in trailing-edges of wings, which unfold as missile leaves container-launcher.
Guidance and Control: Wire guidance. Missile is spin-stabilised by bent-over rear corners of wings. Control by vibrating spoilers.
Warhead: High-explosive, weighing 4.1lb (1.9kg).
Length: 2ft 9½in (0.85m).
Body diameter: 4.3in (11cm).
Wing span: 1ft 3¾in (0.40m).
Launch weight: 16.5lb (7.5kg).
Cruising speed: 190mph (306km/h).
Range limits: 820-6,600ft (250-2,000m).

Development and Service
This very small wire-guided anti-tank missile was designed for operation by a single infantryman. Twelve can be carried in a state of instant readiness by the Puch-Haflinger light cross-country vehicle, each complete with a carrying rack in case they are needed for infantry use in combat areas. Bantams have also been fired successfully from Malmö Mili-trainer light combat aircraft and Agusta-Bell 204 helicopters.

Weight of the complete Bantam weapon system, including launcher, carrying rack, cable and control unit, is 44lb (20kg). Time required to set up the missile and fire it is about 25 seconds. Current contracts from the Swedish and Swiss Armies will maintain production into the mid-seventies, with further orders anticipated.

Bloodhound 2s of the RAF

Bloodhound Mk 2
(Swedish Air Force Designation: RB68)

UK

Surface-to-air missile. In service.

Prime contractor: British Aircraft Corporation, Guided Weapons Division.
Powered by: Two Rolls-Royce Bristol Thor ramjet engines. Four jettisonable Bristol Aerojet solid-propellant boosters.
Airframe: Cylindrical metal body with pointed ogival nose-cone. Pivoted mid-set wings, mid-way back on body and indexed in line with fixed horizontal tail surfaces. Ramjet engines carried on pylons above and below rear part of body. Equi-spaced wrap-round boosters, each with large stabilising fin.
Guidance and Control: Semi-active radar homing, using Ferranti Firelight or GEC/AEI Scorpion target illuminating radar. Twist-and-steer control by means of wings, which can pivot both differentially and together.
Warhead: High-explosive, with proximity fuse.
Max length: 27ft 9in (8.46m).
Max body diameter: 1ft 9½in (54.6cm).
Wing span: 9ft 3½in (2.83m).
Max range: over 50 miles (80km).

Development and Service

Bloodhound was developed originally by The Bristol Aeroplane Co Ltd, under the codename Red Duster, as the standard RAF home defence missile. The original Mk 1 version became the first British weapon in this category to enter service, in mid-1958. It was replaced subsequently by the Mk 2, as described above, with more powerful Thor ramjets and boosters, CW (continuous wave) radar guidance, longer range, greater lethality and improved effectiveness at low altitudes. Tests showed that Bloodhound 2 could destroy fast targets at heights below 1,000ft (300m). Its air transportability was such that it could be used to reinforce overseas commands and, in fact, the missile was deployed by the RAF in Malaysia from the Autumn of 1964. When British forces left the area, these weapons were taken over by the Singapore Air Defence Command. Other operators of Bloodhound 2 are Sweden and Switzerland.

RAF Strike Command continues to deploy Bloodhound Mk 2 in the UK, and No 25 Squadron is operational with similar weapons at three airfields in Germany.

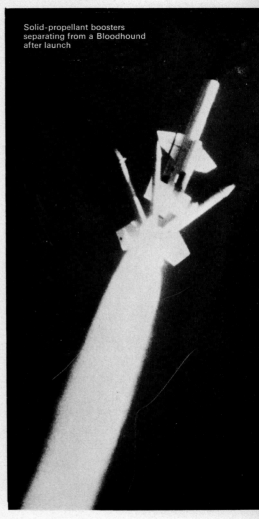

Solid-propellant boosters separating from a Bloodhound after launch

Blowpipe

UK

Shoulder-fired close-range surface-to-air missile. In production.

Prime contractor: Short Brothers and Harland Ltd.
Powered by: Solid-propellant rocket motor.
Airframe: Slim clylindrical body with long ogival nose-cone. Cruciform delta-shape control surfaces on nose, and cruciform delta tail-fins mounted on a sliding ring which locks into position at launch.
Guidance and Control: Radio command guidance. Control by means of foreplanes.
Warhead: High-explosive, with infra-red actuated proximity fuse.
Length: 4ft 5.1in (1.35m).
Body diameter: 3in (7.6cm).
Fin span: 10.8in (27.4cm).
Weight of complete system, with missile: 40lb (18kg).

Development and Service

Shorts initiated development of Blowpipe as a private venture, using a similar command link guidance system to that which had proved so successful with Seacat and Tigercat. Small enough to be carried by a paratroop, fired from the shoulder and airdropped in a multi-round pack, the weapon system consists only of the missile itself in a transport/launch container and an aiming unit attached to the container. All equipment is packed into the aiming unit, including the monocular sight and thumb-operated joystick by which the Blowpipe missile is guided to the target. It can be fired from the ground or from a stationary vehicle, with a reaction time of 20 seconds. As the missile leaves the end of the container, the tail-fin assembly housed in the front of the tube slides along the missile body and locks into place. This enables the diameter of the main part of the container to be kept to a minimum.

Current development of Blowpipe is under Government contract, to meet British Army and Royal Navy requirements for a small anti-aircraft weapon. One application is in the Vickers Shipbuilding Group's SLAM submarine-launched air missile system (see page 136).

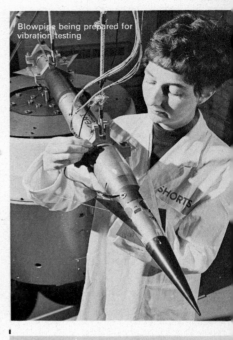

Blowpipe being prepared for vibration testing

Blowpipe leaving its shoulder-fired launcher

Blue Steel

UK

Air-to-surface strategic stand-off missile. In service.

Prime contractor: Hawker Siddeley Dynamics Ltd.
Powered by: Rolls-Royce Bristol Stentor BSSt 1 twin-chamber liquid-propellant rocket engine.
Airframe: Circular-section body, with long pointed ogival nose. Two mid-set delta-shape control surfaces on nose and two vertical delta-shape tail-fins, the upper embodying a rudder and the lower being foldable during take-off under a launch-aircraft, to provide ground clearance. Mid-set wings at tail, each with aileron and anhedral tip.
Guidance and Control: Elliott-Automation inertial guidance. Control by twist and steer, via aerodynamic surfaces.
Warhead: Thermonuclear.
Length: 35ft 0in (10.67m).
Max body diameter: 4ft 2¼in (1.28m).
Wing span: 13ft 0in (3.96m).
Range: approx 200 miles (320km).

Development and Service
This very formidable supersonic thermonuclear missile was the prime instrument of Britain's deterrent policy for many years, as a standard weapon with Vulcan Mk 2 and Victor Mk 2 bombers of the Royal Air Force. Its development was started in 1954 by A V Roe & Co (now embodied in Hawker Siddeley), and the first full-scale round was air-launched in 1958, powered by a de Havilland Double Spectre rocket-engine. This was replaced by the Stentor in production Blue Steels, which entered service with the Vulcan Mk 2s of No 617 Squadron (the wartime dambusters) in the Summer of 1962. No 617 became operational in the following February, and other squadrons of Vulcans and Victors joined the Blue Steel force soon afterwards. As Soviet defence capability improved, the bombers acquired very effective ECM devices, in order to maintain their

Blue Steel on its transporter

Blue Steel under the fuselage of a Vulcan B Mk 2

ability to penetrate to their assigned targets. The final stage, before responsibility for the nuclear deterrent passed to the Royal Navy Polaris submarines, saw both the bombers and the Blue Steels being modified, and the crews being trained, for low-level penetration 'under the enemy radar cover'. By 1972, only Nos 27 and 617 Squadrons, equipped with Vulcans at RAF Scampton, remained operational with Blue Steel.

Bomarc (CIM-10B)

USA

Long-range surface-to-air missile. In service.

Prime contractor: The Boeing Company.
Powered by: Two Marquardt RJ43-MA-7 ramjet engines, each giving 12,000lb (5,440kg) st. Thiokol M-51 solid-propellant integral booster of 50,000lb (22,680kg) st.
Airframe: Aeroplane configuration. Cylindrical metal body with ogival glassfibre radome at nose. Mid-set wings, with pivoted outer panels; tail unit of conventional configuration, with pivoted horizontal surfaces and pivoted top to vertical fin. Ramjets pod-mounted under body on pylons.
Guidance and Control: Initial Western Electric radio command guidance, with active radar terminal homing. Control by aerodynamic surfaces.
Warhead: Nuclear.
Length: 45ft 1in (13.74m).
Body diameter: 2ft 11in (0.89m).
Wing span: 18ft 2in (5.54m).
Launch weight: 16,032lb (7,272kg)
Cruising speed: Mach 2.8.
Max range: 440 miles (700km).

Development and Service

One of the few really long-range surface-to-air weapons yet put into large-scale service, Bomarc dates from a 1949 concept by Boeing (the BO in its name) and the Michigan Aeronautical Research Center (MARC) of the University of Michigan. Starting in 1959, more than 200 test launches were made from Eglin AFB, Florida. These demonstrated that Bomarc could be integrated into America's SAGE (semi-automatic ground environment) defence system, which located potential targets over the North American continent and then directed on to them the aircraft or missiles most likely to ensure their destruction; and that it could intercept supersonic targets at heights far above those at which any contemporary combat aircraft or air-breathing missile operated. The original MIM-10A version of Bomarc became operational in December 1960, but was superseded within five years by the much improved Bomarc B, as described above. This version introduced a solid-propellant booster in place of the former liquid-propellant type, more powerful ramjets and Westinghouse terminal homing effective at all heights from sea level to 100,000ft (30,500m). Reaction time became almost instantaneous, and a single missile site could provide defensive cover for an area of about 500,000 sq miles (1,300,000km^2). On March 23, 1961, a Bomarc B intercepted a Regulus 2 supersonic target drone at a height of 100,000ft, 446 miles (718km) from its launcher. Missiles of this type remain operational at five sites in northeastern USA. Two Canadian squadrons were scheduled to disband in 1972.

Bomarcs elevating from their launch shelters during a practice alert

Bulldog

USA

Tactical air-to-surface missile. Under development.

Development and Service
Under US Navy contract, Texas Instruments Inc is developing guidance and control packages to adapt the Bullpup missile (page 23) into a laser-guided weapon named Bulldog. In operation, the target would be 'illuminated' by a laser device. An electro-optical seeker in the Bulldog would then pick up and home on laser energy reflected from the target. The Naval Weapons Center, China Lake, is responsible for flight tests of the missile.

Bullpup A (AGM-12B/E)

USA

Air-to-surface tactical missile. In service.

Prime contractors: Martin Marietta Corporation/Maxson Electronics Corporation/Kongsberg Vaapenfabrikk.
Powered by: Thiokol LR58-2 storable liquid-propellant rocket motor, of 12,000lb (5,440kg) st.
Airframe: Cylindrical body with pointed ogival nose and tapered tail. Rear-mounted cruciform wings indexed in line with cruciform delta control surfaces on nose-cone.
Guidance and Control: Radio command guidance. Control by foreplanes.
Warhead: High-explosive, weighing 250lb (113kg).
Length: 10ft 6in (3.20m).
Body diameter: 1ft 0in (30cm).
Wing span: 3ft 1in (0.94m).
Launch weight: 571lb (260kg).
Cruising speed: Mach 1.8.
Max range: 7 miles (11km).

Development and Service
Bullpup was conceived originally, under the US Navy designation ASM-N-7, as a simple weapon built around a standard 250lb bomb and powered by a solid-propellant motor developed by the Navy. The pilot of the launch aircraft steered it in flight by movements of a hand switch in the cockpit, using tracking flares above and below the rocket nozzle as a reference in order to keep Bullpup on a line-of-sight path to the target. This original version became operational on April 25, 1959, and was later redesignated AGM-12A. It was superseded by the AGM-2B (formerly ASM-N-7A) for the US Navy, with storable liquid-propellant engine, improved warhead and extended range control; and the USAF AGM-12B (GAM-83A) with a modified guidance system which permitted pilots to make their attacks from an offset position in relation to the target.

Production was undertaken initially by Martin Marietta, who had developed Bullpup A from design study to flight test in under two years. Maxson was brought in as second-source supplier to meet demands for the weapon, and eventually took over all US Bullpup production. Licence manufacture in Europe was handled by a consortium led by AS Kongsberg Vaapenfabrikk of Norway,

Bullpup A missiles on a USAF F-100 Super Sabre

for the armed services of Denmark, Greece, Norway, Turkey and the UK. Aircraft armed with Bullpup have included the US Navy's A-4, A-5, A-6, A-7, F-8E and P-3B; the USAF's F-4, F-100 and F-105, and the Royal Navy's Scimitar and Buccaneer aircraft.

The final version of Bullpup, produced on a limited basis by Martin Marietta for the USAF, was the AGM-12E with an anti-personnel warhead for use in Vietnam.

Bullpup B (AGM-12C/D)

USA

Air-to-surface tactical missile. In service.

Data apply to AGM-12C.
Prime contractor: Martin Marietta Corporation/Maxson Electronics Corporation.
Powered by: Thiokol LR62 storable liquid-propellant rocket motor.
Airframe: Basically-cylindrical body, tapering towards nose and tail. Pointed ogival nose-cone, with cruciform delta control surfaces, mounted on front of body. Rear-mounted cruciform wings indexed in line with foreplanes.
Guidance and Control: Radio command guidance. Control by foreplanes.
Warhead: High-explosive (AGM-12C), or alternative high-explosive or nuclear (AGM-12D).
Length: 13ft 7in (4.14m).
Body diameter: 1ft 6in (45cm).
Wing span: 4ft 0in (1.22m).
Launch weight: 1,785lb (810kg).
Range: 10 miles (16.5km).

Development and Service

Bullpup A proved such an effective weapon that the US Navy sponsored the development of a version with much extended capabilities, built around a 1,000lb (454kg) high-explosive warhead. Known as Bullpup B, this went into production for the Navy under the designation ASM-N-7B, changed after a time to AGM-12C, and augmented rather than replaced Bullpup A. The USAF acquired a version designated AGM-12D (originally GAM-83B), with interchangeable nuclear and high-explosive warheads.

Three Bullpup Bs carried by an A-4 Skyhawk

Chaparral (MIM-72A) USA

Surface-to-air weapon system. In production and service.

Prime contractor: Philco-Ford Corporation, Aeronutronic Division.

Development and Service

Chaparral is a close-range battlefield weapon system which uses Sidewinder 1C missiles (page 134) modified for a surface-to-air role. Four Sidewinders are mounted in ready-to-fire condition on a fire unit which is normally installed on an M730 (modified Army M548 self-propelled tracked cargo vehicle); alternatively the fire unit can be trailer-mounted or dismounted for use from a semi-mobile emplacement. Initial aiming is done by a gunner in the fire unit turret mount, who keeps an optical sight aligned on target. After launch, the missile's own infra-red homing system takes over.

Development of Chaparral began in the Spring of 1965. Test firings from prototype fire units started in July of that year and production was initiated in 1966 for the US Army's then-new air defence battalions, each of which has two batteries of Chaparrals and two of Vulcan multi-barrel 20mm guns. The five-man crew of an M730 have rations and equipment for at least three days of autonomous operations and can travel at up to 40mph (64km/h) on roads. Only one man is required to fire the missiles, of which twelve are carried; the others keep watch for hostile aircraft.

Chinese ICBM

China

Land-based intercontinental ballistic missile. Under development.

Powered by: Liquid-propellant rocket motors.
Warhead: Nuclear.
Range (estimated): 6,000 miles (9,600km).

Development and Service

China achieved its first, successful test firing of an ICBM late in 1970. The range of the missile, which carried an inert warhead, was restricted deliberately to 2,000 miles (3,200km), so that the test could be confined within the nation's own borders; but the eventual production version is expected to have a range of at least three times that distance. Its engines will probably be liquid-propellant rockets, although a factory has already been established to produce solid-propellant motors for second-generation long-range missiles. The US Defense Department has remarked that China is unlikely to begin deployment of ICBMs before 1973-75, and that even then the missiles are not expected to carry penetration aids. Such aids would require a detailed knowledge of the operating characteristics of America's Safeguard ABM system (page 104). An extensive radar and instrumentation system would also be needed for even the simple technique of in-flight fragmentation of the missile's fuel tanks. Nonetheless, it is estimated that China could have a force of 10-25 nuclear-tipped ICBMs operational two or three years after the initial deployment. Twenty-five such missiles launched against the USA with 3-megaton warheads would have the potential to cause 11 or 12 million fatalities in the absence of an efficient ABM system, assuming 40 per cent reliability in reaching their objectives.

Chinese MRBM

China

Medium-range ballistic missile. In production and service.

Warhead: Nuclear.
Range: over 1,000 miles (1,600km).

Development and Service

US Secretary of Defense Melvin R Laird said in February 1970 that China was expected to deploy its first medium-range ballistic missiles (MRBMs) 'at any time' and that 80-100 of the weapons would probably be operational by the mid-seventies. Subsequent reports have suggested that such missiles are emplaced at three operational sites in Tibet, with two more sites under construction in the Winter of 1970-71, some 15,000ft (4,575m) above sea level. The warheads for long-range rockets have been under rapid development since the first Chinese explosion of a 20-kiloton nuclear device at Lop Nor in the Sinkiang desert on October 16, 1964. The fourth such test, in October 1966, was claimed to involve delivery of the warhead to a target area by a missile—thought to have been based on a Soviet weapon like 'Sandal' (page 109). The satellites put into orbit by China in 1970-71 were probably launched by vehicles related to the now-operational MRBM.

Cobra 2000 (BO 810)   Germany

Lightweight anti-tank missile. In production and service.

Prime contractor: Messerschmitt-Bölkow-Blohm GmbH.
Powered by: Solid-propellant rocket motor. Non-jettisonable solid-propellant booster rocket.
Airframe: Cylindrical fibre-paper body, with pointed conical nose. Large plastic cruciform wings, each with spoiler. Booster mounted under body.
Warhead: Hollow charge, weighing 5.5lb (2.5kg) and able to penetrate more than 18.7in (475mm) of steel armour.
Length: 3ft 1½in (0.95m).
Body diameter: 3.9in (10cm).
Wing span: 1ft 7in (0.48m).
Launch weight: 22.5lb (10.2kg).
Max speed: 190mph (306km/h).
Range limits: 1,310-6,560ft (400-2,000m).

Development and Service
This typical one-man anti-tank weapon system comprises the Cobra missile, a control box and cable links; up to eight rounds can be fired selectively from a single control box. No launcher is required, as the missile is supported on the ground by the lid which covers its rear end during transport. The control box is completely self-contained and carries the firing button and control stick.

More than 120,000 Cobras have been delivered to the Danish, German, Italian, Pakistani and Turkish armies, the current version being known as the Cobra 2000 to indicate the extension of its maximum range to 2,000m.

Condor development round on the port underwing launcher of an A-6A Intruder

Condor (AGM-53A)

USA

Tactical air-to-surface missile. Under development.

Prime contractor: North American Rockwell Corporation.
Powered by: Rocketdyne solid-propellant rocket motor.
Airframe: Cylindrical body with rounded glass-tipped nose. Cruciform delta wings, indexed in line with cruciform tail control surfaces.
Guidance and Control: Command guidance from launch aircraft. Operator has TV picture of target area, transmitted via camera in missile nose. Control by tail-fins.
Warhead: High-explosive.
Range: 40-57 miles (64-92km).

Development and Service

Although Condor is a command-guided attack weapon, its long range gives it a stand-off role and the launch aircraft can, in fact, turn for home once the missile has been launched. The TV camera in the nose of the missile sends back a picture of the area towards which it is travelling, and the missile operator is thus able to steer the missile until it locks on to the target. A nominal range of 40 miles (64km) has been quoted, but test firings against land and ship targets, with inert and live warheads, are said to have achieved direct hits over ranges as great as 57 miles (92km) from the launch aircraft. Condor will equip the US Navy's A-6 Intruder and A-7 Corsair II attack aircraft.

Crotale inside its launcher on a Hotchkiss-Brandt AFV, beside the fire-control radar antenna

Crotale/Cactus France

Close-range surface-to-air missile. In production

Prime contractor: SA Engins Matra.
Powered by: Single-stage solid-propellant rocket motor.
Airframe: Slim cylindrical body, with cruciform wings mounted on ogival nose-cone and indexed in line with cruciform tail control surfaces.
Guidance and Control: Radio command guidance. Control by tail surfaces.
Warhead: High-explosive, weighing 33lb (15kg), with infra-red proximity fuse.
Length: 9ft 5¾in (2.89m).
Body diameter: 5.9in (15cm).
Wing span: 1ft 9¼in (0.54m).
Launch weight: 176lb (80kg).
Max speed: Mach 2.3.
Range limits: 1,640ft-5.3 miles (500m-8.5km).

Development and Service

Matra designed this highly-mobile all-weather missile to deal with aircraft flying at speeds up to Mach 1.2 at any height from 165ft (50m) to 9,850ft (3,000m), with a reaction time of only six seconds. First news of the system was given in May 1969, when the South African Defence Minister announced that French companies were developing for his country a surface-to-air weapon known as Cactus. It was revealed subsequently that the prime contractor was Matra and that the same weapon system was known as Crotale in France. Integration with land vehicles was under way by 1969, in which year successful launches were made against target drones. Full production for South Africa and Lebanon was planned for 1972, and the French Government will also take delivery of Crotale systems for airfield defence. The US Army Missile Command is evaluating Crotale as a possible replacement for Chaparral, and Thomson-CSF is adapting the weapon for a ship-based role as part of its Murène system (page 84).

Four Crotale/Cactus missiles are carried in ready-to-fire condition on the standard Hotchkiss-Brandt all-terrain launch vehicle. This carries also a monopulse fire control radar which can track and transmit command signals simultaneously to two missiles. Acquisition after launch is facilitated by an infra-red device which locks on to the missile exhaust. A second vehicle of the same type carries a pulse-Doppler surveillance and target acquisition radar with a range of 11.5 miles (18.5km), and can serve three launch vehicles. Twelve targets can be tracked simultaneously, and 12 missiles can be launched in pairs at six targets in 11 seconds.

Dragon (M47)

USA

Light anti-tank and assault missile. In production.

Prime contractor: McDonnell Douglas Astronautics Company.
Powered by: Several pairs of small solid-propellant rocket motors, in rows around body.
Airframe: Cylindrical body, tapering towards tail and with short ogival nose. Three curved tail-fins which flip open after launch.
Guidance and Control: Wire-guidance, by automatic command-to-line-of-sight system. Control by side-thrusters around body.
Warhead: High-explosive.
Length: 2ft 5.3in (0.74m).
Fin span: 1ft 1in (33cm).
Launch weight: 13.5lb (6.13kg).

Development and Service

Dragon was projected to replace the 90 mm recoilless rifle, to which it is superior in range and accuracy. It was known initially as MAW (Medium Anti-tank/assault Weapon) and is claimed to carry a warhead large enough to destroy most armoured and other infantry targets, while being light enough to be carried and shoulder-fired by one man. In operation, the infantryman first mounts on the missile launch-tube a tracker, embodying the telescopic sight, a sensor device and electronics package. The glass-fibre launch-tube forms a sealed container for the missile during transport and storage, and is enlarged at the aft end to accommodate a propellant container and breech. After acquiring the target in the sight, the operator launches the Dragon missile. As long as the sight remains on the target, the tracker will sense the position of the missile relative to the line of sight and transmit signals to maintain or correct the flight path. 'Steering' is achieved by causing the appropriate pairs of rocket motors, or side thrusters, to fire for both propulsion and guidance.

Development of the Dragon weapon system was started in 1964. Manned shoulder-fired tests began in mid-1968, and service testing in 1971. Dragon was expected to enter full production in the 1972 fiscal year, and will probably equip the US Marine Corps as well as the Army.

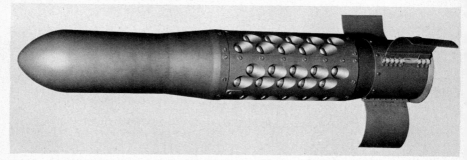

Entac (MGM-32A)   France

Anti-tank missile. In service.

Prime contractor: Nord-Aviation/Aérospatiale, Division Engins Tactiques.
Powered by: Two-stage solid-propellant rocket motor.
Airframe: Cylindrical body with rounded ogival nose. Rear-mounted cruciform wings, each with a spoiler.
Guidance and Control: Wire-guided and roll-stabilised. Control by wing spoilers.
Warhead: High-explosive, weighing 9lb (4.1kg) and able to penetrate more than 25.5 in (650mm) of steel armour.
Length: 2ft 8½in (0.82m).
Body diameter: 6in (15cm).
Wing span: 1ft 3in (0.38m).
Launch weight: 26.9lb (12.2kg).
Max speed: 190mph (305km/h).
Range limits: 1,300-6,600ft (400-2,000m).

Development and Service
One of the first wire-guided anti-tank missiles to enter large-scale production, Entac (ENgin Téléguidé Anti-Char) was adopted in its Model 58 form by the French Army in 1957. A total of 132,000 had been ordered by the armies of 13 countries by early 1971, when the last 3,000 were being manufactured at the rate of 600 a month. Entac is known as the MGM-32A by the US Army.

Entac is intended primarily for use by infantry and was developed by the Direction Technique des Armements Terrestres (DTAT). An operator can control and fire up to ten missiles from a single firing post, with a maximum operator-to-missile distance of 360ft (110m). Alternatively, four Entacs can be mounted in pairs on each side of a jeep, inside their box-like container-launchers. The operator uses an optical sight to locate and track the target and a manual control to command the missile on to his line-of-sight.

Exocet (MM-38) — France

Ship-launched surface-to-surface tactical missile. In production and service.

Prime contractor: Aérospatiale, Division Engins Tactiques.
Powered by: Tandem two-stage solid-propellant motor.
Airframe: Cylindrical body, with pointed ogival nose. Cruciform sweptback wings, indexed in line with cruciform tail control surfaces.
Guidance and Control: Inertial guidance system, associated with TRT FM radio-altimeter, and with active terminal homing radar. Control by tail surfaces.
Warhead: High-explosive, weighing 220lb (100kg).
Length: 16ft 9½in (5.12m).
Body diameter: 1ft 1½in (0.344m).
Wing span: 3ft 3¼in (1.00m).
Launch weight: 1,587lb (720kg).
Cruising speed: High subsonic.
Max range: 23 miles (37km).

Development and Service

Following the sinking of the Israeli destroyer *Eilat* by 'Styx' ship-to-ship missiles, it became a matter of high priority to evolve a defence against such attack. Once missiles like 'Styx' have been launched, they are difficult to intercept. In consequence, Nord-Aviation (now Aérospatiale) concentrated on producing a weapon that would sink enemy missile-armed boats before they were near enough to launch their own attack. Exocet is the answer to that requirement and is one of the most formidable naval weapons ever conceived, being able to travel to the target only 2 to 3 metres (6.5-10ft) above the water, in all weathers, at just below the speed of sound. An oncoming Exocet would thus be extremely difficult to detect by shipboard radar, and it is designed to be impervious to enemy electronic countermeasures (ECM).

Exocet can be fitted to any class of surface warship, including fast patrol boats, provided the ship is fitted with surveillance and target indicating radar, a vertical reference plane gyro and a log to indicate its speed through the water, plus a fire control system. The range and bearing of the target are fed into the missile's guidance system before launch; its cruising height is maintained at very low level by a radio-altimeter linked to its controls. Deliveries were expected to begin in 1972, against initial orders from the Navies of Britain, France, West Germany, Greece and several other countries. Development of Exocet for shore-to-ship and air-to-surface roles is being studied.

Falcon (AIM-4A/C/D/H) USA

Air-to-air missile. In service.

Data apply to AIM-4D
Prime contractor: Hughes Aircraft Company.
Powered by: Thiokol M58-E4 solid-propellant rocket motor of 6,000lb (2,720kg) st.
Airframe: Cylindrical body with tapered and rounded glass-tipped nose. Cruciform vanes aft of nose, indexed in line with long-chord cruciform delta wings. Cruciform tail control surfaces aft of wing trailing-edges.
Guidance and Control: Infra-red homing guidance. Control by tail surfaces.
Warhead: High-explosive.
Length: 6ft 7½in (2.02m).
Body diameter: 6.4in (16.25cm).
Wing span: 1ft 8in (0.51m).
Launch weight: 134lb (61kg).
Max speed: Mach 4.
Max range: 6 miles (9km).

AIM-47A Falcon (*top*) and AIM-4D Falcon (*foreground*)

Development and Service
Hughes began work on the original GAR-1 semi-active radar homing version of Falcon in 1947. The first of about 4,000 production models appeared in 1954, followed by more than 12,000 improved GAR-1Ds (later redesignated AIM-4As). Simultaneously, the company delivered about 16,000 GAR-2s and 9,500 GAR-2As, with pursuit-course infra-red homing. Later redesignated AIM-4C, these offered the advantage that the launch aircraft was free to break off the engagement once the self-homing missiles had been launched. The final production version of this basic Falcon family was the AIM-4D, combining the basic airframe of the 'C' with the improved infra-red homing head of the AIM-4G Super Falcon. Many thousands of older Falcons were converted into 'Ds', which continue to arm F-4 fighters of USAF Tactical Air Command and F-101 and F-102 fighters of Air Defense Command. Under a 1969 USAF contract, Hughes evolved the AIM-4H with new warhead incorporating an active optical proximity fuse (AOPF). A solid-state laser device in the AOPF detonated the warhead without requiring a direct hit, making the 'H' better suited to close-in air combat, but budget economies precluded further development and production. (See also HM-58, page 34, and Super Falcon, page 152).

Falcon (AIM-47A) USA

Air-to-air missile.

Prime contractor: Hughes Aircraft Company.
Powered by: Lockheed Propulsion Company storable liquid-propellant rocket motor.
Airframe: Cylindrical body, with blunt-tipped ogival nose. Long-chord cruciform wings, each with a tail control surface close to its trailing-edge.
Guidance and Control: Infra-red/pulse-Doppler radar guidance. Control by tail surfaces.
Warhead: Interchangeable nuclear or high-explosive.
Length: 12ft 0in (3.66m).
Body diameter: 1ft 1in (33cm).
Launch weight: 800lb (363kg).
Max speed: Mach 6.
Max range: 100 miles (160km).

Development and Service
This very formidable air-to-air weapon was developed originally under the designation GAR-9 to arm the North American F-108 interceptor, which was cancelled. It was resurrected under its present designation as armament for the Lockheed YF-12A experimental Mach 3 fighter, which could carry eight AIM-47As in its internal weapon bays. No production application has yet been announced.

Two RB27 (HM-55) missiles under the fuselage of a J 35F Draken

Falcon (HM-55/HM-58) USA
(Swedish Air Force designations RB27 and RB28)

Air-to-air missiles. In production and service

Prime contractors: Hughes Aircraft Company/Saab-Scania Aktiebolag.
Powered by: Single-stage solid-propellant motor.
Guidance and Control: HM-55 has Hughes semi-active radar homing guidance; HM-58 uses infra-red homing system.
Warhead: High-explosive, with proximity fuse.
Length (RB27): 7ft 1in (2.16m).
Max body diameter (RB27): 11in (28cm).
Wing span (RB27): 2ft 0in (0.61m).
Launch weight: RB27: 262lb (119kg), RB28: 134lb (61kg).
Max speed: Mach 3.
Max range: 6.2 miles (10km).

Development and Service
The HM-55 and HM-58 export versions of the Falcon are built under licence in Sweden by Saab-Scania, under the Swedish Air Force designations RB27 and RB28 respectively. In general appearance the HM-55 resembles the AIM-26A Nuclear Falcon (page 86), but has a conventional warhead. Its all-weather guidance system makes possible attack from any direction, including head-on, and its launch speed makes it compatible with Mach 2 aircraft. The Swedish Air Force deploys it on J 35F Draken interceptors in combination with the HM-58. This is an infra-red pursuit-course weapon generally similar to the AIM-4C Falcon (page 33). The Swiss Air Force has also adopted the HM-55 as standard armament for its Mirage III-S fighters.

Firestreak

UK

Air-to-air missile. In service.

Prime contractor: Hawker Siddeley Dynamics Ltd.
Powered by: Solid-propellant rocket motor.
Airframe: Cylindrical metal body, with cruciform wings mounted more than half-way back from nose and indexed in line with cruciform tail control surfaces. Pointed eight-sided glass nose, with two narrow rings of small windows over infra-red optics on fore-part of body.
Guidance and Control: Infra-red homing guidance. Control by tail-fins.
Warhead: High-explosive, weighing 50lb (22.7kg), with proximity fuse.
Length: 10ft 5½in (3.19m).
Body diameter: 8¾in (22.5cm).
Wing span: 2ft 5½in (0.75m).
Launch weight: 300lb (136kg).
Cruising speed: above Mach 2.
Range limits: 0.75-5 miles (1.2-8km).

Development and Service

Although Firestreak is a comparatively complex and expensive weapon, few other air-to-air missiles can offer such a high single-round kill probability. It is said to have achieved a success rate of more than 85 per cent in trials by the RAF and Royal Navy at all altitudes and in all weather conditions. Control is by a proportional navigation system. After the nose-mounted infra-red unit has brought the missile close to the target, two rings of infra-red optics lock on to the enemy aircraft and feed in angular readings to give bearing and range data during the final stage of interception. Firestreak's only limitation is that it must be fired from astern the target—a shortcoming that does not apply to the further-developed Red Top (page 103). It is standard armament on RN Sea Vixen and RAF Lightning interceptors.

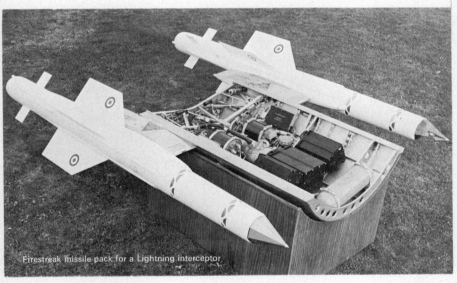

Firestreak missile pack for a Lightning interceptor

Firestreaks on a Lightning of No 56 Squadron, RAF

Frog-1 missile launchers

Frog-1

USSR

Unguided surface-to-surface missile. In service.

Powered by: Solid-propellant rocket motor, with seven nozzles.
Airframe: Single-stage cylindrical metal structure, with six fixed tail-fins and large, bulbous warhead at nose.
Guidance and Control: No guidance system. Spin stabilised.
Warhead: Alternative nuclear and high-explosive types.
Length: 31ft 0in (9.50m).
Fin span: 3ft 3in (1.0m).
Launch weight: 6,000lb (2,700kg).
Max range: 15 miles (24km).

Development and Service
One of the first Soviet missiles revealed in a Moscow parade, this artillery rocket has been in large-scale service with the Soviet and other Warsaw Pact armies for at least 15 years. Its NATO code-name is an acronym of 'Free Rocket Over Ground', indicating that no guidance system is fitted.

'Frog-1' can be regarded as Russia's counterpart to Honest John (page 54). Its tracked transporter is based on the standard Soviet JS-3 amphibious reconnaissance vehicle chassis. The ribbed casing in which the missile is housed elevates to form the launch-tube in action. Soviet commentators have referred to 'Frog-1's' high precision and enormously powerful thermonuclear warhead.

Frog-2, 3, 4, and 5

USSR

Unguided surface-to-surface bombardment missiles. In service.

Data apply to Frog-4.
Powered by: Two-stage solid-propellant motors. Twelve small nozzles around main nozzle of booster.
Airframe: Tandem two-stage all-metal casings, of cylindrical shape. Conical nose-cone. Fixed cruciform tail-fins.
Guidance and Control: No guidance system. Spin stabilised.
Warhead: Alternative nuclear or high-explosive.
Length: 33ft 6in (10.20m).
Fin span: 3ft 6in (1.05m).
Launch weight: 4,400lb (2,000kg).
Max range: 30 miles (50km).

Development and Service
Like 'Frog-1', these missiles are all unguided artillery rockets; but each has two stages in tandem and is carried on an uncovered launch rail by a tracked vehicle based on the PT76 amphibious reconnaissance tank. The original 'Frog-2', of which eight were paraded through Moscow in November 1957, had a bulbous warhead and was about 29ft 6in (9.0m) long. It was followed successively by 'Frog-3', with a more cylindrical bulbous warhead; the 'Frog-4', differing only in having a warhead of the same diameter as the second-stage casing; and 'Frog-5' with yet another variation in warhead. These rockets were all delivered in large numbers to the Soviet Army and its Warsaw Pact allies.

Frog-5 transporter/launcher on exercise

Frog-5 launch crew in protective clothing after leaving a 'contaminated area' during nuclear warfare training

Frog-7 weapon system

Frog-7

USSR

Unguided surface-to-surface bombardment missile. In service.

Powered by: Solid-propellant motor. Eighteen small nozzles around main nozzle.
Airframe: Cylindrical casing, with cruciform cropped-delta tail-fins. Pointed nose-cone.
Guidance and Control: No guidance system. Spin stabilised.
Warhead: Alternative nuclear or high-explosive.
Range limits: 4.5-57 miles (7.5-92km).

Development and Service

This latest known version of the Frog (Free Rocket Over Ground) series was first seen in Moscow on November 7, 1965. It differs from 'Frogs 2-5' in several important respects, notably in being a single-stage weapon. Also, like most of the other battlefield missiles identified since the mid-sixties, it is transported on a wheeled vehicle rather than the former tracked launchers.

'Frog-7s' have been deployed rapidly in the Soviet Army and with the armed forces of Russia's friends and allies, including the Warsaw Pact nations. Large numbers were reported in Egypt in 1970-71. The latest available photographs show the transporter/launcher carrying two additional missiles, to each side of the ready-to-fire 'Frog-7' on the launch-rail.

Frog-7 missile unit under training

Gabriel — Israel

Ship-launched surface-to-surface missile. In production and service.

Prime contractor: Israel Aircraft Industries Ltd.
Powered by: Unspecified rocket motor.
Airframe: Basically cylindrical body, tapering towards the tail. Rectangular cruciform wings and tail surfaces, indexed in line with each other. Ogival nose-cone.
Guidance and Control: Advanced electronic guidance system, with terminal homing.
Warhead: High-explosive, weighing 330lb (150kg).
Length: 11ft 0in (3.35m).
Body diameter: 12.8in (32.5cm).
Wing span: 4ft 6½in (1.39m).
Launch weight: 882lb (400kg).
Max range: 12.6 miles (20km).

Development and Service

This subsonic ship-to-ship weapon is the only missile of Israeli design of which details have yet been released officially. Its obvious sophistication and capability lend emphasis to reports of more advanced weapons such as the MD-660 (see page 78). Gabriel was developed by IAI in collaboration with the Israeli Navy, the obvious purpose being to provide a counter to missile-launching ships like the Egyptian *Osa* class which sank the Israeli destroyer *Eilat* in 1967. It has been operational since early 1970 on Israel's French-built *Saar* class gunboats, each of which carries two three-round swivelling launchers on the after-deck and two single-round launchers, on each side of a 40-mm gun, on the forward deck.

Gabriel is delivered inside a glass-fibre container which embodies the launching rail and is simply installed on the ship-board mountings without further test or adjustment. It can be launched in bad weather or rough seas, and is said to approach its target only a few metres above the water, lessening the possibility of detection. IAI describe it as an automatic homing weapon and claim that it is impervious to enemy electronic counter-measures.

Gainful (SA-6)

Surface-to-air missile. In service.

Powered by: Solid-propellant rocket motor.
Airframe: Cylindrical body, with cruciform wings at mid-body, indexed in line with cruciform tail surfaces. Slim cylindrical duct from forward of wings almost to tail on side of body. Shorter ducts of similar form above and below body.
Length: 19ft 6in (6.0m).

USSR

Development and Service
First displayed in public during a Moscow parade on November 7, 1967, this compact anti-aircraft missile has since become a standard weapon in the Soviet forces and is reported to be deployed in Egypt. It is somewhat larger than the American Hawk (page 52) but almost certainly fulfils a similar role, by providing a rapid-reaction defence against aircraft flying at low and medium altitudes. Three 'Gainful' missiles are carried side-by-side atop a tracked transporter/launcher.

Galosh   USSR

Anti-ballistic missile. In service.

Powered by: Rocket motors with four first-stage nozzles.

Development and Service

'Galosh' has been under development at least as long as its US counterparts of the Safeguard ABM system. The ribbed cylindrical container in which it is transported (and from which it is, presumably, fired) was first displayed on a tractor/trailer in a Moscow parade on November 7, 1964. On that occasion the Soviet commentator referred to 'Galosh' as an anti-ballistic-missile (ABM) weapon. More recent American official statements have suggested that 'Galosh' is the missile installed at 64 ABM launch sites around Moscow, each with very advanced early warning radars and tracking equipment. One radar, known to NATO as 'Hen House', is described as being 'the size of three football fields lined up end-to-end and standing on their sides ... providing the same radar coverage that the US will have by 1978 if all of the Safeguard programme is completed.'

Nothing is known of the 'Galosh' missile itself, except that four first-stage nozzles have been visible inside the container, which is 67ft (20.4m) long with a diameter of 9ft (2.75m). A newsreel film shown in Moscow in 1965 claimed to depict an ABM launching, and may have portrayed an early 'Galosh' test round. This appeared to have a somewhat similar shape to the American Sprint (page 140) with less taper on the main body and a conventional conical nose-cone. It suggested that the missile breaks through the spherical end-cap on the launch-tube when it is fired.

Ganef (SA-4)　　　　　　　　　　　　　　USSR

Surface-to-air missile. In service.

Powered by: Ramjet sustainer. Four solid-propellant wrap-round boosters, with canted nozzles, which jettison after burn-out.
Airframe: Cylindrical main body, with smaller-diameter centre-body carrying pointed conical warhead. Cruciform pivoted wings on fore-part of main body, indexed at 45 degrees to fixed cruciform tail-fins. Boosters equi-spaced around body aft of wings.
Guidance and Control: Command guidance system. Control via pivoted wings.
Warhead: High-explosive.
Length: 30ft 0in (9.15m).
Max body diameter: 2ft 8in (0.80m).
Wing span: 7ft 6in (2.30m).
Ceiling: 80,000ft (24,400m).

Development and Service
This weapon was probably evolved to provide Soviet land forces with a highly mobile anti-aircraft defence effective up to the service ceiling of most contemporary combat aircraft. The annular air-intake between the main body and the centre-body suggests the use of a ramjet sustainer engine, which would be consistent with long range. The twin round tracked launcher carrying 'Ganef' was first shown in a Moscow parade in May 1964. Since then this missile system has been deployed widely in the Soviet Union and is reported to have been despatched to Egypt as an element of the Soviet forces supporting that nation. 'Ganef' can be transported on its launch vehicle by the An-22 freight aircraft. It may have a surface-to-surface capability.

Genie (AIR-2A) USA

Unguided air-to-air nuclear missile. In production and service.

Prime contractor: McDonnell Douglas Astronautics Company.
Powered by: Thiokol SR49-TC-1 solid-propellant rocket motor of 36,000lb (16,330kg) st.
Airframe: Cylindrical body with cruciform tail-fins, each with movable tip, and an ogival nose of greater basic diameter than body.
Guidance and Control: No guidance system. Trajectory stabilised by fins and gyroscope.
Warhead: Nuclear, with yield reported as 1.5 kilotons.
Length: 9ft 7in (2.91m).
Body diameter: 1ft 5.35in (0.44m).
Fin span: 3ft 3¼in (1.00m).
Launch weight: 820lb (372kg).
Max speed: Mach 3.
Max range: 6 miles (9.6km).

Development and Service

Although Genie is an unguided rocket, it justifies inclusion in this book by reason of its nuclear warhead. It was the first nuclear-tipped air-to-air missile ever tested in a live firing when, on July 19, 1957, it was launched from a Northrop F-89J Scorpion at Indian Springs, Nevada. Immediately after launch, at a height of 15,000ft, the pilot of the fighter turned sharply to escape the blast and the missile was detonated from the ground after travelling 3 miles (4.8km). USAF observers who stood directly under the point of detonation for an hour suffered no ill effects.

Many thousands of Genies have been delivered and continue in first-line service with F-101B and F-106 squadrons of the USAF and CF-101F squadrons of the Canadian Armed Forces. They are designed to be fired automatically and detonated by the Hughes fire-control system fitted in each aircraft. A training version, without nuclear warhead, is also in service.

Goa (SA-3/SA-N-1)　　　　　　　　　　　　　　　　　USSR

Surface-to-air close-range missile. In service.

Powered by: Two-stage rocket motors. Booster is solid-propellant.
Airframe: Two basically-cylindrical stages, in tandem. Booster fitted with large rectangular cruciform fins. Second stage has cruciform fixed wings at rear and cruciform cropped-delta control surfaces on the tapered nose-cone.
Guidance and control: Control by movable foreplanes.
Warhead: High-explosive.
Length: 19ft 4in (5.90m).
Body diameter: Booster 2ft 3in (0.70m). 2nd stage 1ft 6in (0.45m).
Wing span: 4ft 0in (1.22m).
Max range: 15 miles (24km).

Development and Service

Russia's early counterpart to the American Hawk, 'Goa' is the anti-aircraft missile which caused much concern to the Israeli Air Force when 14 batteries supplemented earlier 'Guidelines' near the Suez Canal by early 1971. It is effective over a slant range of about 15 miles (24km), to a height of 40,000ft (12,200m), and is known in America as the SA-3 or SAM-3 in its land-based version. Its compact form is evident from the fact that pairs of 'Goas' can be transported on the truck that is used as a tractor for the 'Guideline' and 'Guild' trailer-transporters.

In its ship-based form (SA-N-1), 'Goa' is standard armament in the Soviet Navy, on a twin-round launcher very like that used for the American Tartar missile. Six cruisers of the *Kresta* class and 15 destroyers of the *Kashin* class each carry two launchers, fore and aft. Four cruisers of the *Kynda* class have a single twin-launcher on their fore-deck; three destroyers of the *Kanin* class and seven of the *Kotlin* class have a similar launcher on their after deck.

A photograph of a twin-launcher on a destroyer, issued by *Tass* in 1971, showed missiles of the 'Goa' type with an additional set of small tail-fins between the booster and second-stage wings. The wings also appear to carry trailing-edge control surfaces. These missiles may be similar to the new SAM (SA-N-3) carried by the helicopter cruisers *Leningrad* and *Moskva*

Goas on a mobile land-based launch vehicle

Typical ship-borne Goa twin-launcher. These missiles are of the latest type with additional tail-fins

Griffon — USSR

Long-range surface-to-air missile. In service.

Powered by: Multi-stage rocket motors. Solid-propellant first-stage booster.
Airframe: Cylindrical booster, with large cruciform cropped-delta fins. Smaller-diameter second stage is also cylindrical, with ogival nose; it carries large cropped-delta cruciform wings, each with a control surface inset in its trailing-edge. Cruciform tail control surfaces, on tapering tail of second stage, are indexed in line with wings and at 45 degrees to booster fins. Warhead may separate as third stage and use an in-built rocket motor during final approach to target.
Guidance and Control: Radar command. Control by wing trailing-edge surfaces and second-stage tail-fins.
Warhead: Probably interchangeable nuclear and high-explosive types.
Length (with nose-probe): 54ft 0in (16.50m).
Body diameter: Booster 3ft 6in (1.07m).
2nd stage 2ft 10in (0.85m).
Wing span: 12ft 0in (3.65m).

Development and Service
Largest Russian surface-to-air missile yet revealed publicly, 'Griffon' has been trundled through Moscow on a trailer regularly since November 1963. It was then described by the Soviet commentator as an anti-missile missile, but does not appear to have the capability of intercepting an ICBM warhead. Its use is almost certainly limited to an anti-aircraft role, with limited effectiveness against tactical ballistic missiles and air-to-surface weapons.

Guideline (V750VK) (SA-2/SA-N-2) USSR

Surface-to-air missile. In service.

Data apply to 'Mk 2' version.
Powered by: Liquid-propellant sustainer; solid-propellant booster.
Airframe: Tandem two-stage missile. Booster is cylindrical with cruciform cropped-delta fins, two of which have trailing-edge control surfaces. Smaller-diameter second stage has cylindrical body with flared skirt, fixed cruciform wings about mid-way between nose and tail, and small cruciform tail control surfaces, all indexed in line with booster fins. Small cruciform vanes on nose.
Guidance and Control: Automatic radio command. Control by movable tail surfaces and control surfaces on booster fins.
Warhead: High-explosive, weighing 288lb (131kg) with contact or proximity fuse, or detonated by command.
Length: 34ft 9in (10.60m).
Body diameter: Booster 2ft 2in (0.66m).
2nd stage 1ft 8in (0.51m).
Wing span: 5ft 7in (1.70m).
Launch weight: 5,000lb (2,270kg).
Max speed: Mach 3.5.
Max range: 25 miles (40km).

Development and Service
Most widely used of all Soviet anti-aircraft missiles, 'Guideline' was first displayed in Moscow in 1957. It serves throughout the Soviet Union and the Warsaw Pact countries, and has been supplied in vast numbers to Cuba, Egypt, Indonesia, Iraq, North Vietnam and other countries. At least 28 batteries were reported to be operational near the Suez Canal in the Spring of 1971.

'Guidelines' captured by Israeli forces during the 1967 war were said to bear the Soviet designation V750VK, with the designation V75SM applicable to the entire weapon system. This includes a Zil 157 semi-trailer transporter-erector vehicle, radar van and generators. The US designation is SA-2, or SAM-2; while the version carried on a twin-launcher by the Soviet Navy cruiser *Dzerzhinski* is known as SA-N-2. 'Guideline' has undergone considerable development. The original version, which might be referred to for convenience as the 'Mk 1', was similar to the first-generation US Nike-Ajax in many design features and probably had a comparable performance. Initially the nose vanes were rectangular, but the 'Mk 2' version captured by the Israelis had cropped-delta vanes. The latest 'Mk 4' version, first displayed in Moscow in 1967, is about 15in (40cm) longer than the 'Mk 2', and lacks the latter's nose vanes and booster control surfaces. It was described by the Soviet commentator as being far more effective than earlier versions, which may imply use of a nuclear warhead in the bulged, white-painted nose-cone.

Three 'Mk 4' Guideline missiles (*foreground*)

V75SM weapon system, comprising Guideline missiles, radar van and generator.

SA-N-2 Guidelines on the Soviet cruiser *Dzerzhinski*

Guild   USSR

Surface-to-air missile. In service.

Powered by: Solid-propellant rocket motor.
Airframe: Cylindrical body, with pointed ogival nose. Cruciform cropped-delta wings, each with trailing-edge control surface. Cruciform foreplane control surfaces indexed in line with wings.
Guidance and Control: Control by foreplanes and wing trailing-edge surfaces.
Warhead: High-explosive.
Length: 39ft 0in (12.0m).

Development and Service
'Guild' was the second type of Soviet surface-to-air missile displayed in a Moscow parade, on November 7, 1960. Unlike 'Guideline' it has no separate booster, which may imply the use of a dual-thrust solid-propellant motor. Although there is good reason to believe that 'Guild' is a standard defensive weapon in the USSR, there is no evidence to suggest that it has been exported.

Hardsite   USA

Anti-ballistic missile defence system. Under development.

Development and Service
The US Army began advanced development of prototype Hardsite Defense (HSD) components in the 1971 Fiscal Year. No details may yet be published; but HSD is envisaged as an 'add-on' to the Safeguard ABM system (see page 104), enabling it to cope with new Soviet submarine-launched ballistic missiles and land-based ICBMs fitted with multiple independently-targeted re-entry vehicles (MIRV). Elements of HSD include a new computer radar and a new high-acceleration short-range missile to supplement Sprint.

HARM   USA

High-velocity air-to-surface anti-radar missile. Under development.

Prime contractor: Naval Weapons Center, China Lake, California.

Development and Service
Little information on this new missile is yet available. Its development was prompted by experience in Vietnam, where Soviet-built radars were often able to detect oncoming first-generation anti-radiation weapons such as Shrike and shut down before the missile could home on their emissions. HARM (High-velocity Anti-Radiation Missile) is intended to have such a high performance that it would home on any radar before the operators had time to switch off.

Harpon   France

Anti-tank missile. In production.

Prime contractor: Nord-Aviation/Aérospatiale, Division Engins Tactiques.
Guidance and Control: Wire-guidance, using TCA optical aiming/infra-red tracking system.

Development and Service.
Harpon is externally similar to the SS.11 B.1 anti-tank missile (see page 145), with similar performance and alternative warheads, but utilises the TCA type of automatic guidance described under the AS.30 entry (page 13). It was, in fact, the first of a long series of weapons which Nord-Aviation (now Aérospatiale) adapted or designed to use this guidance system. Harpon has proved able to operate effectively over ranges as small as 1,200ft (400m) as no time is lost by the operator having to acquire the missile after launch. It is normally deployed on a launcher turret carried by light armoured vehicles, and is claimed to offer a defence against head-on attack by low-flying fixed-wing aircraft or helicopters in a surface-to-air role.

Harpoon in surface-launched form, with solid-propellant booster attached in tandem

Harpoon (ZAGM-84A) USA

Air-to-surface and surface-to-surface anti-shipping missile. Under development.

Prime contractor: McDonnell Douglas Corporation.
Powered by: Either Garrett AiResearch ETJ 331 or Teledyne CAE turbojet engine. Solid propellant booster for surface to surface launch only.
Airframe: Slim cylindrical body, with ogival nose. Cruciform wings and cruciform tail surfaces, indexed in line. Vertical air intake for turbojet. When tandem booster is fitted, its cruciform fins are indexed in line with missile fins.
Dimensions, Weight and Performance: Secret.

Development and Service

After three years of research and study, and evaluation of competing tenders from five major US aerospace companies, McDonnell Douglas was named prime contractor for this important weapon in June 1971. Its major sub-contractor is Texas Instruments Inc, and the programme is being managed by the US Naval Air Systems Command, with support from Naval Ordnance Systems Command. This reflects the fact that Harpoon is intended to be launched from both aircraft and ships, against shipping targets, over extended stand-off ranges. Few details are available, except that the missile will be turbojet-powered. Mock-up missiles have been shown on underwing launchers on F-4 Phantom aircraft and housed in an Asroc eight-round launcher. Use of such existing launchers is expected to save money and speed entry of the Harpoon into service. The present contract covers demonstration of test versions of the missile over a two-year period. If all goes well, it will be followed by a final development contract and start of production in about 1975.

Hawk (MIM-23A)

USA

Surface-to-air close-range missile. In production and service.

Data apply to Basic Hawk.
Prime contractor: Raytheon Company.
Powered by: Aerojet-General M22 E8 dual-thrust solid-propellant motor.
Airframe: Cylindrical metal body with pointed ogival nose. Long-chord cruciform wings, each with trailing-edge elevon.
Guidance and Control: Continuous-wave semi-active radar homing system by Raytheon.
Warhead: High-explosive.
Length: 16ft 6in (5.03m).
Body diameter: 1ft 2in (0.36m).
Wing span: 3ft 11.4 in (1.20m).
Launch weight: 1,295lb (587kg).
Max speed: Mach 2.5.
Max range: 22 miles (35km).
Ceiling: over 38,000ft (11,600m).

Development and Service

Manufactured in thousands, in Europe and Japan as well as by Raytheon in America, Hawk (Homing-All-the-Way Killer) is the low/medium-altitude anti-aircraft weapon which set the pace in this difficult role. As early as January 1960 a Hawk intercepted and destroyed an Honest John supersonic artillery rocket, so becoming the first missile to bring down a ballistic missile. Later, Hawks achieved equal successes against a smaller Littlejohn and Corporal guided ballistic missile. They became operational in August 1960 and were subsequently deployed by the US Army and Marine Corps in Europe, Korea, Okinawa, the Panama Canal Zone and Vietnam. Other nations equipped with Hawks include Belgium,

Hawk self-propelled launcher

Hawk tracked loader, with launcher in background

France, Israel, Italy, Japan, Korea, Netherlands, Saudi Arabia, Spain, Sweden, Taiwan and West Germany.

The complete Hawk weapon system is transportable by fixed-wing aircraft and helicopters. A typical battery consists of six three-round launchers, a tracked loader which collects three missiles at a time for delivery to the launchers, a pulse acquisition radar, a CW acquisition radar, a range-only radar, two target illumination radars and a battery control centre. The original trailer-type launcher has been replaced in some US Army units by a tracked, self-propelled launcher able to tow an item of ground support equipment. The effectiveness of Hawk, down to treetop level, results from the fact that its radars can pick out reflected signals from a moving target at low altitude from the mass of signals reflected by objects on the ground.

In 1964, Raytheon began development of Improved Hawk, a completely redesigned missile with solid-state guidance package, larger warhead and improved Aerojet-General motor propellant. This was approved for full production, as a follow-on to Basic Hawk, in early 1971.

Hellcat UK

Air-to-surface tactical missile.

Prime contractor: Short Brothers and Harland Ltd.

Development and Service
Hellcat is a projected air-to-surface version of Seacat (page 121), which could be mounted on most types of military helicopter. It is designed to use standard Seacat missiles, in conjunction with a thumb-operated controller and gyro-stabilised sight in the cockpit of the launch aircraft.

HOBOS USA

Homing bomb system. In production and service.

Prime contractor: North American Rockwell Corporation.
Airframe: Cylindrical nose and tail assemblies, linked by four long-chord, short-span strakes, fit over standard bomb casing. Cruciform tail-fins integral with rear end of strakes, each fitted with trailing-edge flap control surface.
Guidance and Control: Various types of guidance system can be installed. Current HOBOS bombs utilise TV system as described under entry for Hornet (page 00). Control by flaps on tail surfaces.
Warhead: Current HOBOS missiles are built around standard MK 84 (2,000lb; 907kg) and M118E1 (3,000lb; 1,360kg) bombs.

Development and Service
HOBOS (HOming BOmb System) comprises a kit of parts which can be mounted on a variety of standard unpowered bombs to convert them into highly-accurate guided weapons. Each kit consists of a forward guidance assembly, the interconnect section (including the bomb) and an aft control section (including an autopilot). Most current production HOBOS missiles are built around MK 84 bombs and use the TV type of guidance that was developed on the Hornet experimental air-to-surface missile. The normal bomb suspension and release systems are unchanged.

Honest John (MGR-1) USA

Surface-to-surface unguided artillery missile. In service.

Data apply to MGR-1B
Prime contractor: US Army Missile Command (USAMICOM).
Powered by: Hercules M-31A1 single-stage solid-propellant rocket motor.
Airframe: Cylindrical metal body, with bulged ogival nose-cone and four clipped-delta tail-fins.
Guidance and Control: Spin stabilisation only; initial spin imparted by Thiokol rockets aft of warhead and maintained throughout flight by canted tail-fins.
Warhead: Alternative nuclear or high-explosive, weighing about 1,500lb (680kg).
Length: 24ft 9¼in (7.55m).
Body diameter: 2ft 6in (0.76m).
Fin span: 4ft 6in (1.37m).
Launch weight: 4,719lb (2,140kg).
Max speed: Mach 1.5.
Range limits: 4.6-23 miles (7.4-37km).

Development and Service
Although unguided, this big artillery rocket has provided powerful nuclear support for NATO and Allied armies for two decades. Its development was undertaken by Douglas, under the technical supervision of the US Army's Ordnance Missile Laboratories at Redstone Arsenal, Alabama. Firing tests began in 1951, only one year after receipt of the initial contract, and the original MGR-1A version of Honest John was manufactured subsequently in very large numbers by both Douglas and Emerson Electric.

Continued development led to the MGR-1B, reduced in length by about 2ft 5½in (0.76m) but with improved performance. This version entered first-line service in 1960 and is still deployed by US Army units in Europe and Japan, as well as by other countries such as Britain, Italy, France and Germany. The US Marine Corps formed three Honest John batteries, but disbanded them when the weapon system proved too cumbersome for amphibious operations.

Hornet (ZAGM-64A)   USA

Air-to-surface and surface-to-surface anti-tank missile. Under development.

Prime contractor: North American Rockwell Corporation.
Powered by: Dual-thrust solid-propellant rocket motor.
Airframe: Cylindrical body; rear-mounted cruciform wings, each with a trailing-edge flap control surface. TV homing version has hemispherical glass nose.
Guidance and Control: Variety of alternative guidance systems. Version illustrated uses TV guidance. Control by flaps on wings.
Warhead: None at present stage of development.
Range: more than 2.3 miles (3.7km).

Development and Service

North American Rockwell's Columbus Division developed Hornet originally under a USAF contract, in 1965-66, to demonstrate the feasibility of TV homing guidance for an air-to-surface anti-tank missile. Launches from an F-100 resulted in five direct hits on stationary and moving targets, but Hornet was not developed further. Instead, the USAF adapted the Hornet guidance system for the HOBOS guided bomb (page 53).

In 1970, the US Army decided to utilise a new version of Hornet to test a variety of laser and TV seekers that were projected for the next generation of close-support missiles. A larger rocket motor was fitted for ground launching, and the control system was modified to use improved cold-gas actuators; otherwise the current version—known as a Terminal Homing Flight Test Vehicle—is little changed. When using the TV version, the operator has only to lock the missile's stabilised vidicon camera on to the target before launch, by means of a TV monitor. The missile then homes automatically on the target after launch, without further action by the operator.

HOT

France/Germany

Anti-tank missile. Under development.

Prime contractors: Aérospatiale, Division Engins Tactiques/Messerschmitt-Bölkow-Blohm GmbH.
Powered by: Two-stage solid-propellant rocket motor.
Airframe: Basically cylindrical body, with aft section of increased diameter and bulged ogival nose. Four flip-out fins.
Guidance and Control: Wire guidance, using TCA optical aiming/infra-red tracking system. Spin-stabilised by fins. Control by vanes in rocket exhaust.
Length: 4ft 2in (1.27m).
Body diameter: 5.5in (14cm).
Fin span: 1ft 0¼in (0.31m).
Launch weight: 44lb (20kg).
Max speed: 625mph (1,010km/h).
Range limits: 250-13,100ft (75-4,000m).

Development and Service

HOT (High-subsonic Optically-guided Tube-launched) is similar in configuration to MILAN (see page 79) and uses the same guidance system, but is larger and has a high performance. It is intended to be carried in two three-round packs, mounted on each side of the turret of light armoured vehicles such as the French AMX-13 tank and the West German SPZ. During 1970 demonstration firings, two HOTs from an AMX-13 impacted within 12in (30cm) of the centre of a target over a range of 11,480ft (3,500m). Others were fired over a similar range from an Alouette III helicopter, and the prime contractors are developing an automatic guidance system for such air-to-surface use of HOT. They claim that this weapon, like MILAN, has potential also in a surface-to-air role against head-on attack by helicopters and low-flying fixed-wing aircraft. Production of the system is dependent on the outcome of competitive evaluation against the French ACRA and American TOW.

Hound Dog (AGM-28) USA

Air-to-surface strategic stand-off missile. In service.

Prime contractor: North American Aviation Inc.
Powered by: Pratt & Whitney J52-P-3 turbojet, rated at 7,500lb (3,400kg) st.
Airframe: Cylindrical body, with movable delta foreplanes on long pointed ogival nose. Rear-mounted delta wings, with ailerons, and vertical tail-fin and rudder. Engine pod-mounted beneath rear of body, on a short pylon.
Guidance and Control: Inertial guidance system by North American Autonetics, supplemented by a star-tracking system supplied by Kollsman Instrument Company. Control by aerodynamic surfaces.
Warhead: Thermonuclear, reported yield 4 megatons.
Length: 42ft 6in (12.95m).
Body diameter: 2ft 4½in (0.72m).
Wing span: 12ft 2in (3.71m).
Launch weight: 9,600lb (4,350kg).
Cruising speed: Mach 2.
Max range: 600 miles (965 km).

Development and Service

America's counterpart to the British Blue Steel, Hound Dog differs in being turbojet-powered which gives it a longer range. Its development was started in 1957, to provide primary armament for the B-52G/H versions of the Stratofortress strategic bomber, which carries two Hound Dogs on underwing pylons. The weapon system entered service on the 'G' in December 1959. Earlier versions of the B-52 were subsequently adapted to carry and launch Hound Dog, and it was deployed with 29 Strategic Air Command wings by August 1963. Several hundred missiles remained operational in the early 'seventies, with regular refurbishing to maintain their capability. Most are of the AGM-28B (formerly GAM-77A) version, with modifications to improve navigational accuracy.

Hound Dog is stored and mounted on its launch aircraft with its wing pylon in place. This pylon contains the astro-tracking system, which supplements data fed into Hound Dog's inertial navigation system from the B-52's navigation system before launch. The missiles' engines can be run during take-off of the heavily-laden launch-aircraft, to shorten its run. This does not shorten the missiles' range, as their tanks can be topped up in flight from those of the bomber. The aircrew can also make changes of target and missile flight profile whilst airborne, before the missiles are launched. Ceiling of the Hound Dog is given as above 52,000ft (15,850m).

Ikara

Australia

Long-range anti-submarine missile. In production and service.

Prime contractor: Department of Supply, Commonwealth of Australia.
Powered by: Dual-thrust solid-propellant rocket motor.
Airframe: Basically square-section body, with mid-mounted cropped-delta wings, elevon control surfaces and upper and lower tail-fins. Torpedo cradled in underbody recess.
Guidance and Control: Radio/radar guidance system, receiving signals from automatic target-prediction equipment on board launch-ship. Control via elevons.
Warhead: American Type 44 lightweight acoustical homing torpedo, or other types of torpedo.
Length: 11ft 0in (3.35m).
Wing span: 5ft 0in (1.50m).
Weight and Performance: Secret.

Development and Service

This long-range anti-submarine missile was designed originally to provide ships of the Royal Australian Navy with a rapid-reaction weapon which would guarantee a high degree of accuracy under all weather conditions. Three destroyers of the *Perth* class are each fitted with two single-round launchers; the six *River* class destroyers each have one launcher. A version known as RN Ikara is being developed jointly by the Australian Department of Supply and Hawker Siddeley Dynamics of the UK for the Royal Navy. This will be fitted on RN *Leander* class ships, using a launch system of Vickers design, and will also be carried by the new guided missile destroyer HMS *Bristol*.

Ikara missiles, with torpedoes already in place, are stowed in a shipboard magazine, from which they can be moved rapidly on to the swivelling launcher. When an enemy submarine is detected by sonar on board the launch ship, other surface ships or helicopters, the launcher is turned towards the target and the missile is fired. It is tracked and guided from the launch ship during its flight. At the optimum position for attack, the torpedo is released by separating the ventral tail-fin assembly and is lowered into the water by parachute. It then homes on to sound emissions from the enemy craft.

Ikara launcher

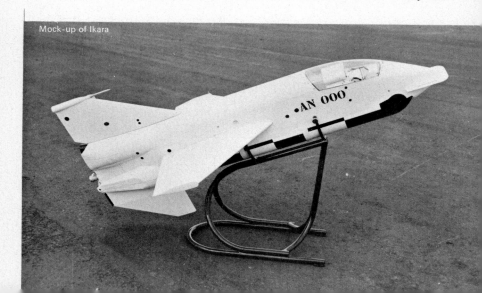

Mock-up of Ikara

Indigo

Italy

Surface-to-air short-range missile. Under development.

Prime contractor: Sistel—Sistemi Elettronici SpA.
Powered by: Solid-propellant rocket motor of 7,055lb (3,200kg) st.
Airframe: Cylindrical light alloy body with pointed ogival nose. Movable cruciform wings mid-way along body, indexed in line with fixed cruciform tail-fins.
Guidance and Control: Beam-riding/radio command guidance. Control by movable cruciform wings.
Warhead: High-explosive warhead of fragmentation type, with Hawker Siddeley Dynamics infra-red proximity fuse.
Length: 10ft 6in (3.20m).
Body diameter: 7.5in (19cm).
Wing span: 2ft 7in (0.79m).
Launch weight: 214lb (97kg).
Max speed: Mach 2.5.
Range limits: 0.6-6.2 miles (1-10km).

Development and Service
Following successful firing trials at the Italian range in Sardinia, Indigo is being evaluated by the Italian Army. Contraves Italiana, one of the parent companies of Sistel, has developed an integration kit enabling the missile to be operated by anti-aircraft gun batteries equipped with either the Super Fledermaus fire control system designed by Contraves AG of Switzerland or the CT40-G system of its own design. Beam-riding guidance is standard, with stand-by radio command and infra-red tracking to counter enemy jamming or ECM.

KAM-3D (Type 64 ATM-1)   Japan

Anti-tank missile. In production and service.

Prime contractor: Kawasaki Jukogyo Kabushiki Kaisha.
Powered by: Dual-thrust solid-propellant rocket motor, first stage rated at 286lb (130kg) st and second stage at 33lb (15kg) st.
Airframe: Cylindrical steel body, with a blunt-top conical centre-body on the rounded nose. Large rear-mounted cruciform wings, of all-metal construction, each with full-span trailing-edge spoiler.
Guidance and Control: Wire guidance, with gyro-stabilisation. Control by spoilers on wings.
Warhead: High-explosive.
Length: 3ft 4in (1.02m).
Body diameter: 4.7in (12cm).
Wing span: 1ft 11½in (0.60m).
Launch weight: 34.6lb (15.7kg).
Cruising speed: 190mph (306km/h).
Range limits: 1,150–5,900ft (350–1,800m).

Development and Service

Kawasaki started development of this missile in 1957, under contract to the Technical Research and Development Institute of the Japan Defence Agency. Seven years later, after several hundred rounds had been test fired, production was authorised for the Japan Ground Self-Defence Force, under the official designation Type 64 ATM. Experience has shown that the KAM-3D's velocity control system is so effective that three out of four unskilled operators can hit the target with their first round after simulator training. Skilled operators can achieve a 95 per cent success rate. A two-man firing team is standard, using a push-button control box. The KAM-3D can be fired by infantry as single rounds or from multiple-round emplacements. It is also carried on jeeps and helicopters. Optical tracking is aided by a flare in daytime; at night the rocket exhaust provides adequate visual reference.

KAM-3D on Bell 47 helicopter

KAM-9 (TAN-SSM) Japan

Anti-tank missile. Under development.

Prime contractor: Kawasaki Jukogyo Kabushiki Kaisha.
Powered by: Daicel dual-thrust solid-propellant sustainer. Nippon Oils and Fats Co solid-propellant rocket booster.
Guidance and Control: Nippon Electric Co wire-guidance system, with optical aiming and automatic tracking.
Warhead: Armour-piercing type.
Length: 4ft 11in (1.50m).
Body diameter: 5.9in (15cm).
Wing span: 1ft 1in (0.33m).

Development and Service
Under development since 1966, this missile is described as a higher-performance, extended-range version of the KAM-3D, suitable for use against armoured vehicles on both land and water. It differs from the earlier missile in several important respects. It is, for example, launched from the tubular container in which it is transported and stored. In action, the container is placed on a launch and tracking unit, incorporating the firing mechanism, optical sight and missile check-out device. A booster fires briefly to eject the KAM-9 from the launch-tube. Then, at a safe distance, the sustainer fires to accelerate the missile to cruising speed. The operator has only to keep his optical sight aligned on the target. Deviations from the line-of-sight track to the target are sensed automatically and corrective signals are fed from a computer to the missile via the wire-guidance system. Pre-series production of the KAM-9 was started in 1970.

Kangaroo (AS-3) USSR

Air-to-surface strategic missile. In service.

Powered by: Unidentified turbojet engine.
Airframe: Aeroplane configuration. Sweptback wings and tail unit, with vertical surfaces of rhomboid form. Nose air-intake.
Guidance and Control: Aerodynamic control surfaces.
Length: 48ft 11in (14.9m).
Max range: 400 miles (650km).

Development and Service
'Kangaroo' is similar in size and configuration to a single-seat, single-engined sweptwing fighter aircraft. It was first displayed, clamped under the cutaway bomb-bay of its Tupolev Tu-95 mother-plane, during the fly-past at the 1961 Soviet Aviation Day display at Tushino Airport, Moscow. Further details became available subsequently when a Soviet film showed a 'Kangaroo' being released from a Tu-95 in flight. In particular, the film showed the shape of the vertical tail surfaces for the first time, and confirmed that the ram air-intake of 'Kangaroo' is blanked off by a duct fairing under the Tu-95's belly in flight. This may convey compressed air from an auxiliary power unit to start the missile's engine before launch. A large radar scanner in the Tu-95's nose is probably used to compute the course to target, so that 'Kangaroo' can be directed initially on the correct compass bearing.

Kangaroo under Tu-95

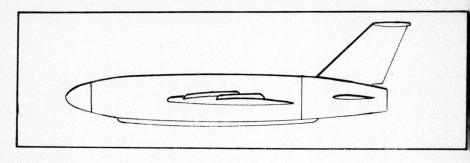

Kelt (AS-5)

Air-to-surface stand-off missile. In service.

Powered by: Unidentified type of rocket motor.
Airframe: Aeroplane configuration, basically similar to that of 'Kennel', with sweptback wings and vertical tail surfaces. Rounded nose radome. Fairing under fuselage similar in form to that of 'Styx'.
Guidance and Control: Possibly radar-homing. Aerodynamic control surfaces.

Development and Service
The underwing pylon on which 'Kelt' is carried by the Tu-16 bomber appears to be more massive than that used for 'Kennel (see below-). This suggests that 'Kelt' is a heavier weapon than its turbojet-powered predecessor. It may also offer a higher degree of accuracy if the much larger nose radome houses a correspondingly large scanner. Published reports imply that 'Kelt' can be carried as a standard anti-shipping missile by Tu-16s of the Soviet Naval Air Arm operating over areas like the North Sea and Mediterranean, from bases in Egypt. Its range is estimated to be more than 100 miles (160km) and it could clearly be used against land bases as well as at sea.

Kennel (AS-1)

USSR

Air-to-surface anti-shipping missile. In service.

Powered by: Unidentified turbojet engine.
Airframe: Aeroplane configuration. Body of circular section, with hemispherical radome above chin air-intake. Mid-set sweptback wings, each with two boundary-layer fences, and sweptback tail surfaces. Tailplane mounted part-way up fin, which has a pod fairing at the tip.
Guidance and Control: Possibly radio command with terminal homing. Control by aerodynamic surfaces.
Warhead: High-explosive.
Length: 27ft 0in (8.2m).
Wing span: 16ft 0in (4.9m).
Max range: 63 miles (100km).

Development and Service
Looking rather like a scaled-down MiG-15 jet-fighter, this anti-shipping missile is thought to have been superseded by the rocket-powered 'Kelt' (see page 62) in Soviet Naval Air service. However, Tupolev Tu-16 ('Badger-B') maritime patrol-bombers of the Indonesian and Egyptian Air Forces were supplied complete with underwing pylons to carry two 'Kennels', and a surface-to-surface version is operational, with the NATO designation 'Samlet' (see page 108).

Kennels under the wings of a Tu-16 of the Indonesian Air Force

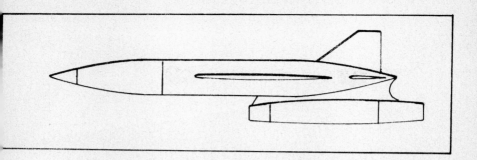

Kipper (AS-2)   USSR

Air-to-surface stand-off missile. In service.

Powered by: Unidentified turbojet engine.
Airframe: Aeroplane configuration, with cylindrical body, ogival nose-cone, mid-set sweptback wings and conventional sweptback tail surfaces. Engine slung under rear fuselage in pod.
Length: 31ft 0in (9.5m).
Max range: 132 miles (213km).

Development and Service
Like 'Kangaroo' and 'Kitchen', this anti-shipping missile put in a surprise first appearance in the 1961 Soviet Aviation Day display at Tushino Airport, Moscow. It was seen to be similar in general configuration to the American Hound Dog, but smaller and of less refined form. 'Kipper' is carried by Tu-16 twin-jet bombers of the Soviet Naval Air Arm, clasped under the modified bomb-bay. This version of the Tu-16 is known to NATO as 'Badger-C'. It differs from the 'Kennel' armed 'Badger-B' in having radar in a wide nose radome, presumably for detecting and fixing the position of the target before 'Kipper' is launched.

'Badger-Cs' have been reported over the North Sea and over the Mediterranean, the latter from bases in Egypt. 'Kipper' could clearly be used against targets on land as well as at sea.

Kitchen (AS-4)   USSR

Air-to-surface strategic missile. In service.

Airframe: Aeroplane configuration. Basically-cylindrical body, with short-span delta wings and cruciform tail surfaces.
Length: 37ft 0in (11.3m).
Max range: 460 miles (740km).

Development and Service
Although this weapon was first identified more than a decade ago, at the 1961 Soviet Aviation Day display, little is known about it. The commentator at Tushino Airport then referred to the Tu-22 carrying 'Kitchen' as the spearhead of the Soviet strategic rocket force, and most of the 22 Tu-22s which took part in the 1967 display at Domodedovo were armed with this missile. There is, however, reason to believe that the Tu-22 failed to fulfil Soviet requirements as a supersonic strategic bomber and has been switched largely to a maritime reconnaissance role. It remains to be seen whether 'Kitchen', or a development of it, will be carried by the Tupolev swing-wing bomber ('Backfire') which is intended to replace the Tu-22 in the Soviet Air Force. Meanwhile, 'Kitchen' remains the most advanced Soviet air-to-surface missile yet known.

Kormoran on an F-1

Kormoran

Germany

**Air-to-surface anti-shipping missile.
Under development.**

Prime contractor: Messerschmitt-Bölkow-Blohm GmbH.
Powered by: Solid-propellant rocket motor, with boost and sustainer stages.
Airframe: Cylindrical body with pointed nose-cone. Cruciform sweptback wings, indexed in line with cruciform tail control surfaces.
Guidance and Control: Inertial guidance system, with active or passive terminal homing by CSF RE576 radar homing head. Control via tail surfaces.
Warhead: High-explosive.
Length: 14ft 5in (4.40m).
Wing span: 3ft 3½in (1.00m).
Body diameter: 1ft 1½in (34.2cm).
Launch weight: 1,280lb (580kg).
Max speed: Mach 0.95.
Max range: 23 miles (37km).

Development and Service

Largest post-war missile project undertaken in West Germany, Kormoran is being developed to equip F-104G Starfighters of the German Naval Air Service. Work on the weapon was started in 1964, initially as a collaborative programme by MBB and Nord-Aviation of France, who allocated the designation AS.34 to the missile. Nord subsequently withdrew, and MBB have continued the development and testing of Kormoran by themselves. The first air launch from an F-104G was made successfully on March 19, 1970. Tests of fully-equipped but inert missiles against targets many miles away from the launch aircraft began with equal success in early 1971. Normally, Kormoran is launched at low level and travels towards the target at an altitude which puts it below shipboard radar cover. At a predetermined distance from the target, it enters a climb to enable its homing radar to 'see' and lock on to the target.

Bloodhound Mk 2 surface-to-air missiles of the Royal Air Force, standing guard near a Strike Command base in Britain

F-8 Crusader fighter weapon test aircraft, carrying two Bullpup A air-to-surface rockets under its wings and Zuni unguided rockets on its fuselage pylons

Two Hound Dog thermonuclear air-to-surface missiles under the wings of a B-52 Stratofortress bomber of USAF Strategic Air Command

F-104G fighter used for development testing of the Kormoran air-to-surface missile. Launchings are photographed by the camera pack under its fuselage

AS.37 anti-radar version of the Anglo-French Martel under the fuselage of a Mirage III-E

Lance combat ballistic missile on its lightweight mobile launcher. This is helicopter-transportable and can be air-dropped by parachute

Roland twin-launcher, with integral surveillance and search radar, mounted on a 25-ton SPZ tank of the Federal German Army

Test firing a Sergeant medium-range field artillery missile. The operational version can be fitted with alternative nuclear or high-explosive warheads

Launching a Shrike air-to-surface missile from an A-4 Skyhawk attack aircraft of the US Navy. Shrike homes on the radars of enemy early warning and missile sites

Three Sparrow air-to-air missiles under the wings and fuselage of an F-4E Phantom II fighter. The port forward missile attachment carries an Eros electronic collision-avoidance system

This dramatic photograph, taken from a circling aircraft, shows two Spartan anti-ballistic missiles on their way to intercept an ICBM warhead, from Kwajalein Atoll in the Pacific

Bell's advanced KingCobra combat helicopter, carrying a four-round TOW missile pack under each of its stub-wings

Lance/Extended Range Lance (XMGM-52B/XRL)
USA

Surface-to-surface artillery missile. In production.

Prime contractor: LTV Aerospace Corporation.
Powered by: Rocketdyne dual-thrust storable liquid-propellant (UDMH and IRFNA) rocket engine. Improved Rocketdyne liquid-propellant engine in XRL.
Airframe: Cylindrical body, with small cruciform tail-fins and ogival nose-cone.
Guidance and Control: Simplified inertial guidance system, developed by US Army Missile Command.
Warhead: Alternative nuclear or high-explosive.
Length: 20ft 0in (6.10m).
Body diameter: 1ft 10in (0.56m).
Launch weight: 2,850lb (1,293kg).
Range limits: 3-30 miles (4.8-48km).

Development and Service

Known originally as Missile B, Lance has been under development since 1962 as a replacement for the US Army's Sergeant and Honest John. It is a divisional support weapon, intended to be transported normally on an XM-667 tracked erector/launch vehicle produced by FMC Corporation, with a further vehicle of the same type to carry two spare missiles and a reloading hoist. Alternatively, Lance can be fired from a lightweight wheeled launcher, produced by Orenda in Canada, which is helicopter-transportable and can be air-dropped by parachute. Each launcher is manned by a six-man firing crew.

Firing trials began in March 1965, leading up to successful climatic trials in Alaska in 1969. Production was deferred to take advantage of improved performance offered by the slightly enlarged Extended Range Lance (XRL), which is now the standard version. It uses the same guidance system and ground equipment as earlier models but is said to offer an increase of at least 80 per cent in range. The Rocketdyne engine is interesting in that it consists of two concentrically-mounted stages. The high-thrust outer stage operates only as a launch booster. The inner stage fires simultaneously, but continues operation as the sustainer, at reduced thrust, when the booster has burned out.

Lance production began under a US Army contract awarded in January 1971, with initial delivery of the complete weapon system for evaluation three months later.

Mace (CGM-13)   USA

Surface-to-surface cruise missile. In service.

Data apply to CGM-13C.
Prime contractor: Martin Marietta Corporation.
Powered by: Allison J33-A-41 turbojet, rated at 5,200lb (2,360kg) st. Thiokol jettisonable solid-propellant booster, rated at 100,000lb (45,360kg) st.
Airframe: Basically cylindrical body, tapered at tail and with rounded nose. Sweptback wings and T-tail. Engine air intake and blister fairing under body.
Guidance and Control: AC Spark Plug AChiever inertial guidance system. Aerodynamic control surfaces.
Warhead: Nuclear.
Length: 44ft 0in (13.40m).
Body diameter: 4ft 6in (1.37m).
Wing span: 22ft 11in (6.98m).
Launch weight: 18,000lb (8,165kg).
Max speed: over 650mph (1,045km/h).
Max range: 1,200 miles (1,930km).

Development and Service

Although the US Services experimented with their first pilotless 'flying bombs' during the 1914-18 War, they had nothing like the German V-1 in the second World War. Impressed by the potential of the enemy V-weapons, they resumed missile development in 1944, and the USAAF decided to concentrate at first on subsonic 'flying bombs', believing that its experience in operating winged aircraft would guarantee better results than were achievable with somewhat-inaccurate rockets of the V-2 type. First product of this policy was the TM-61 Matador, which entered service in 1951 and became the first remotely-controlled pilotless weapon to be deployed overseas when two USAF squadrons were sent to Germany in 1954.

Because Matador's radio-navigation guidance system was susceptible to jamming, Martin evolved from it two improved missiles, the MGM-13B (formerly TM-61B and TM-76A) Mace and CGM-13C (TM-76B) Mace, with self-contained, unjammable guidance systems. The latter continued to equip two 16-missile squadrons of the Pacific Air Forces (PACAF) on Okinawa and one USAFE squadron in Germany in 1971. In each case the missiles are housed in 'hardened' underground launch sites.

Malafon Mk 2 on board the French destroyer *Vauquelin*

Malafon   France

Surface-to-surface or surface-to-underwater missile. In production and service.

Prime contractor: Société Industrielle d'Aviation Latécoère.
Powered by: Unpowered in cruising flight. Launched by two jettisonable solid-propellant boosters which burn for three seconds.
Airframe: Aeroplane configuration. Cylindrical body, with tapered tail-cone and housing for torpedo in nose. Short pivoting wings. High-set tailplane with endplate fins. Wingtip tracking flares.
Guidance and Control: Command guidance, via ship's sonar. 'Twist-and-steer' control by means of pivoting wings.
Warhead: Acoustic-homing torpedo of 21in (53cm) diameter, weighing 1,157lb (525kg). This is ejected by inertia when a tail-parachute is streamed half-a-mile from the predicted position of the target.
Length: 19ft 8in (6.0m).
Wing span: 9ft 10in (3.0m).
Launch weight: 2,865lb (1,300kg).
Max range: 11 miles (18km).
Max speed (at booster burn-out): 515mph (830km/h).

Development and Service

Development of this anti-submarine weapon system began in 1956. Initial launch trials, from the ground and from aircraft, underlined the problem of maintaining a correct cruising height after the launch-boosters had burned out. This was overcome eventually by using a radio altimeter to measure and maintain a constant altitude of about 330ft (100m) by changing the wing incidence to increase lift as the speed of the missile fell. Launch and guidance trials at sea began from the anti-submarine ship *La Galissonnière* in June 1962. The operational sonar system and artillery-type radar were installed on this ship two years later, and operational testing of production Lat-233 Malafon Mk 2 missiles was undertaken in the first half of 1965. Subsequently, launchers have been installed on the frigates *Suffren* and *Duquesne* (each of which carries a single launcher and 13 missiles), five destroyers of the *Surcouf* class and the frigate *Aconit*. Malafon will also arm the three new Type 67 frigates of the French Navy.

Martel (AS-37/AJ.168)

France/UK

Air-to-surface missile. In production.

Prime contractors: SA Engins Matra (France) and Hawker Siddeley Dynamics Ltd (UK).
Powered by: Solid-propellant rocket motors by Hotchkiss-Brandt and Aérospatiale.
Airframe: Cylindrical body, with sweptback cruciform wings, indexed in line with cruciform tail control surfaces. AJ.168 version has a hemispherical glass nose; the AS.37 has a pointed ogival nose.
Guidance and Control: AS.37 homes automatically on emissions from enemy radars. AJ.168 follows a pre-programmed course initially. Final impact is effected by the weapon operator, who is given a visual picture of the target area on a high-brightness monitor by means of a TV camera in the nose of the missile. By movement of a joystick or similar control in the cockpit of the launch aircraft, the operator aligns the TV camera on target. The missile control system then causes the Martel to align its axis with that of the TV camera. Control via tail-fins.
Warhead: High-explosive.
Length: AS.37 13ft 1¼in (4.00m).
AJ.168 12ft 0in (3.65m).
Body diameter: 1ft 3in (0.38m).
Wing span: 3ft 8in (1.12m).
Weight and Performance: Secret.

Development and Service
This very important tactical missile is being developed in two forms for the RAF and French Services. The AJ.168, which will be carried by the RAF's Buccaneer aircraft, is a command guided weapon for stand-off attacks on surface targets and is described as having a range of 'tens of miles'. The French AS.37 is an all-weather weapon for the specific task of destroying enemy radars, on which it homes automatically after launch at very low, medium or high altitudes. In other respects the two versions are identical, and both are able to operate regardless of enemy electronic countermeasures (ECM) activity. The AS.37 will arm the Mirage III-E, Jaguar and Atlantic aircraft of the French Air Force and Navy.

AS.37 Martel

AJ.168 Martel

Masurca Mk 2

France

Ship-based surface-to-air missile. In service.

Prime contractor: Direction Technique des Constructions Navales.

Powered by: Tandem two-stage solid-propellant rocket motors supplied by the Direction des Poudres. Polka booster develops 75,000lb (34,000kg) st; second stage develops 4,585lb (2,080kg) st in the Mod 2 missile, 4,785lb (2,170kg) st in the Mod 3.

Airframe: Tandem two-stage missile. Cylindrical missile body with pointed ogival nose; cruciform wings of long chord and constant narrow span, indexed in line with pivoted cruciform tail control surfaces. Cylindrical jettisonable booster of larger diameter than missile, with cruciform tail-fins indexed in line with those of missile.

Guidance and Control: Mod 2 version has beam-riding guidance and Mod 3 has semi-active radar homing, in each case by CFTH/CSF. Control by tail surfaces.

Warhead: High-explosive, weighing 105lb (48kg), with proximity fuse.

Length: with booster 28ft 2½in (8.60m). without booster 17ft 4½in (5.29m).

Body diameter: 1ft 4in (40.6cm).

Wing span: 2ft 6in (0.77m).

Launch weight (Mod 3): 4,585lb (2,080kg).

Max range (Mod 3): 31 miles (50km).

Development and Service

Developed and produced by the Naval Arsenal at Ruelle, this anti-aircraft missile is carried on a twin-launcher by the French Navy frigates *Suffren* and *Duquesne,* whose main duty is to provide escort for the carriers *Foch* and *Clémenceau.* A similar launcher is being fitted on the cruiser *Colbert* during an extensive refit. Two missiles can be launched in a space of a few seconds against different targets. Each of the frigates carries a total of 48 missiles.

Maverick (*foreground*) on triple underwing launcher

Maverick (AGM-65A) — USA

Tactical air-to-surface missile. In production.

Prime contractor: Hughes Aircraft Company.
Powered by: Thiokol TX-481 solid-propellant rocket motor.
Airframe: Cylindrical body, with rounded glass nose. Long-chord delta wings, indexed in line with cruciform tail control surfaces, mounted close to their trailing-edges.
Guidance and Control: Self-homing TV guidance system. Control via tail-fins.
Warhead: High-explosive.
Length: 8ft 1in (2.46m).
Body diameter: 1ft 0in (30cm).
Wing span: 2ft 4in (0.71m).
Launch weight: under 500lb (227kg).
Performance: Secret.

Development and Service
Development of Maverick began in July 1968, under contract from the USAF Aeronautical Systems Division, to provide a 'shoot and scoot' missile for aircraft like the F-4 Phantom and A-7 Corsair II. The first guided test flight, only 18 months later, achieved a direct hit on a tank target at Holloman range. A $69.9 million initial production contract was awarded to Hughes in late 1971, and deliveries were scheduled to begin in late 1972. In action, Maverick's nose TV camera is focused on target by the pilot of the launch aircraft by means of a monitor screen in the cockpit. After launch, the missile's TV 'eye' remains locked on to the target, guiding it automatically to impact on the precise spot at which the camera is aimed. Meanwhile, the pilot is free to seek out and attack other targets or turn away from the target area. Smallest of several US TV-guided weapons, Maverick is intended for use against pinpoint targets such as tanks and columns of vehicles, as well as field fortifications, radar sites and buildings. Test launches have been made at distances ranging from a few thousand feet to many miles, and from high altitudes down to treetop level. It will normally be carried in two three-round underwing clusters.

MD-660 — Israel

Surface-to-surface bombardment missile.

Prime contractor: Israel Aircraft Industries Ltd.
Powered by: Solid-propellant motors.
Airframe: Two-stage missile.
Warhead: Probably alternative nuclear or high-explosive.
Airframe, Dimensions and Weights: Secret.
Max range: 280 miles (450km).

Development and Service
In view of the delicate situation in the Middle East, it is hardly surprising that no details have come officially from Israeli sources to confirm reports of the existence of this mobile, ramp-launched weapon. The availability of a rocket able to hit major targets throughout the Arab world from inside Israeli territory would be a powerful deterrent factor against future breaches of the peace. Its potential is made infinitely greater by the likelihood that Israel possesses the capability of manufacturing nuclear warheads at the Dimona reactor centre near the Dead Sea.

The designation MD-660 implies that the early development of this missile system was undertaken by Avions Marcel Dassault of France on behalf of the Israeli Government. This may be linked with reports of a Dassault surface-to-surface weapon named Jericho, some years ago, and the knowledge that firing trials of such a missile were being conducted in the Mediterranean, off Toulon, in the Spring of 1968.

MILAN

France/Germany

Infantry anti-tank missile. In production.

Prime contractors: Aérospatiale, Division Engins Tactiques/Messerschmitt-Bölkow-Blohm GmbH.
Powered by: Dual-thrust solid-propellant rocket motor.
Airframe: Basically cylindrical body, with fore and aft sections of larger diameter than centre portion. Four flip-out tail-fins.
Guidance and Control: Wire guidance, using TCA optical aiming/infra-red tracking system. Spin-stabilised by tail-fins. Control by jet-deflection.
Warhead: Shaped charge.
Length: 2ft 5.5in (0.75m).
Body diameter: 4.6in (11.6cm).
Fin span: 10.5in (0.27m).
Launch weight: 13.9lb (6.3kg).
Max speed: 400mph (640km/h).
Range limits: 83-6,560ft (25-2,000m).

Development and Service

MILAN (Missile d'Infanterie Léger ANtichar) is one of three modern-generation battlefield weapon systems being developed jointly by Aérospatiale and MBB, the others being HOT and Roland. It is typical of many current missiles in being fired from its tranport container, which is simply placed on a lightweight launch and guidance unit when required for use. The latter unit embodies the firing mechanism, an optical sight with 4x magnification, an infra-red goniometer and the guidance electronics; it weighs 35lb (16kg), compared with 24.2lb (11kg) for the missile in its container. One man can carry the complete weapon system, accompanied by a colleague carrying two spare missiles.

Compared with the missiles it will replace, MILAN is a high-speed weapon, reducing the opportunity for effective counter-fire from the target. Its initial acceleration is 750g, compared with 15g for first-generation wire-guided missiles. Guidance is by the TCA system, as described for Roland (see page 104). Over short ranges, MILAN can be used as a recoilless rifle, and claims have been made for its effectiveness as a surface-to-air weapon to protect ground targets against head-on attack by helicopters and low-flying fixed-wing aircraft. Development testing of the basic weapon system had been completed by 1971.

Minuteman III launch

Minuteman (LGM-30) USA

Intercontinental ballistic missile. In production and service.

Data: LGM-30G Minuteman III.
Prime contractor: The Boeing Company.
Powered by: First stage: Thiokol M-55E solid-propellant motor. Second and third stages: Aerojet-General solid-propellant motors.
Airframe: Three-stage cylindrical missile. Two upper stages have smaller diameter than first stage. Ogival nose fairing covers warhead. No fins or wings.
Guidance and Control: Inertial guidance system by Autonetics Division of North American Rockwell. First-stage control by four swivelling nozzles.
Warhead: Thermonuclear, in multiple individually-targetable re-entry vehicles (MIRV).
Length: 59ft 10in (18.20m).
Max body diameter: 6ft 0in (1.83m).
Launch weight: 76,000lb (34,475kg).
Max speed: 15,000mph (24,000km/h).
Max range: over 8,000 miles (13,000km).

Development and Service

Minuteman was designed as a second-generation ICBM to supersede the earlier liquid-propellant Atlas and Titan. Its solid-propellant motors enable it to be stored at instant readiness in below-ground silo launchers. Although much smaller and lighter in weight than the first-generation weapons, its range is similar. The warhead is smaller, but the use of MIRV warheads on Minuteman III increases enormously the possibility of penetrating enemy defence systems.

Three versions of Minuteman are currently operational. Wing II at Ellsworth AFB, S Dakota, and Wing V at Warren AFB, Wyoming, have a total of 350 LGM-30B Minuteman Is. These are 55ft 11in (16.99m) long, with weight of 65,000lb (29,500kg) and range of 6,300 miles (10,130km). Each successive stage is smaller than the one before, with a still-smaller cylindrical warhead of more than one megaton in an Avco Mk 11 re-entry vehicle. Wings I at Malmstrom AFB, Montana, III at Minot AFB, N Dakota, IV at Whiteman AFB, Missouri, and VI at Grand Forks, Utah, have LGM-30F Minuteman IIs, of similar configuration to the LGM-30B but 59 ft 10in (18.20m) long, with a weight of 70,000lb (31,750kg), range of 7,000 miles (11,265km) and warhead of over two megatons in a General Electric Mk 12 re-entry vehicle. The first Minuteman III squadron (data above) was installed at Minot in December 1970. Eventually, the force will comprise 490 LGM-30Fs and 510 LGM-30Gs.

The 200 silo launchers of Wing I are dispersed over an area of 18,000 sq miles (46,600km^2). Each silo is 80ft (24.4m) deep. Two Strategic Air Command officers control each flight of 10 missiles from an underground launch centre. Airborne control can be exercised from KC-135 command post aircraft.

Minuteman III

Mosquito

Switzerland/Italy

Lightweight anti-tank missile. In service.

Prime contractor: Contraves Italiana SpA.
Powered by: Two-stage solid-propellant rocket motor.
Airframe: Cylindrical glass-fibre body with pointed conical nose. Folding cruciform wings of sandwich construction, each with trailing-edge vibrating spoiler.
Guidance and Control: Wire guidance, with roll-stabilisation by powder-driven gyro. Control by vibrating spoilers on wings.
Warhead: Hollow charge, weighing 9lb (4kg) and able to penetrate more than 26in (660mm) of armour, or fragmentation type.
Length: 3ft 7.7in (1.11m).
Body diameter: 4.72in (12cm).
Wing span: 1ft 11.6in (0.60m).
Launch weight: 31lb (14.1kg).
Cruising speed: 205mph (330km/h).
Range limits: 1,200-7,800ft (360-2,375m).

Development and Service

This simple, one-man, infantry anti-tank weapon was developed by Contraves-Oerlikon of Switzerland but manufactured until 1971 by Contraves Italiana. It is standard equipment in the Swiss and Italian Armies, who use also a version fitted with a parachute recovery system, instead of a warhead, for training. Six Mosquitos can be carried in their container-launchers, ready for firing, by the Puch-Haflinger light cross-country vehicle. Mosquitos have also been fired successfully from Agusta-Bell 47 helicopters.

In conventional infantry use, the Mosquito is controlled and fired by means of a small control box carrying a joystick and optical sight. The container-launcher houses the missile with its warhead detached, and weighs 48.5lb (22.0kg).

MSBS

France

Submarine-launched medium-range ballistic missile. In production and service.

Prime contractor: Aérospatiale, Division des Systèmes Balistiques et Spatiaux.
Powered by: First stage: PNSM P.10 (Type 904) solid-propellant motor. Second stage: PNSM P.4 (Rita) solid-propellant motor.
Airframe: Two-stage missile. Cylindrical body of constant diameter, with conical nose-cone. First-stage casing of Vascojet 1000; second-stage casing of glass-fibre. No fins or wings.
Guidance and Control: Inertial guidance system. First-stage control by four gimballed nozzles; second-stage control by thrust-vectoring in single fixed nozzle.
Warhead: Nuclear, approx 500-kiloton yield.
Length: 34ft 1½in (10.40m).
Max body diameter: 4ft 11in (1.50m).
Launch weight: 39,683lb (18,000kg).
Performance: Secret.

Development and Service

Already operational on the French Navy's SNLE submarine *Redoutable*, the MSBS (Mer-Sol Balistique Stratégique) will go to sea also on the *Terrible* and *Foudroyant* in 1973-75. Construction of a fourth ship, *l'Indomptable*, is expected to begin in 1976, and a fifth may follow later. Each carries 16 missiles, in an arrangement similar to that of America's Polaris/Poseidon submarines, and is nuclear-powered. A full salvo of 16 missiles could be launched in 15 minutes.

Early firing trials were made from the experimental submarine *Gymnote*, using M-112 test vehicles with a live first stage and dummy second stage. Next came M-012 rounds with two live stages and an inert re-entry system. These were followed in turn by M-013 MSBS prototypes, which became operational on the *Gymnote* before the complete production weapon system could be cleared for operational deployment on the *Redoutable*. Like Polaris/Poseidon, the MSBS is 'popped' from its launch-tube by a separate charge, so that its first-stage rocket motor can ignite immediately after emersion from the water.

Nuclear tests in the Pacific had enabled the French to miniaturise their thermonuclear charges from 2.6 to one megaton by 1970. It is expected that further development will enable the MSBS to be retrofitted with a thermonuclear warhead by 1975.

Murene
France
Ship-based surface-to-air close-range weapon system. Under development.

Prime contractor: Thomson-CSF.

Development and Service
Matra's Crotale all-weather surface-to-air weapon system (see page 29) offers such promise that Thomson-CSF decided to develop a ship-based version as a private venture. The result is Murène, which comprises one or more eight-round turret-launchers, a Triton air and sea surveillance radar with a range of 23 miles (37km), a pulse-Doppler radar providing aerial surveillance and IFF interrogation over an 11.2-mile (18-km) range, Pollux stabilised fire control radar with a range of 10 miles (16km), a digital computer for automatic tracking and a display console. The effectiveness of Murène can be extended by linking it with a Mureca weapon system, which adds a 48-round Catulle rocket-launcher to the Crotale launcher.

Nike Ajax (MIM-3)
USA
Surface-to-air missile. In service.

Prime contractor: Western Electric Company Inc.
Powered by: Tandem Aerojet-General liquid-propellant sustainer of 2,600lb (1,180kg) st and Hercules Inc solid-propellant jettisonable booster, giving 59,000lb (26,760kg) st for 2½ seconds.
Airframe: Two-stage design. Cylindrical missile body, tapering towards long ogival nose, with four small pivoted foreplanes on nose indexed in line with cruciform delta wings at rear. Cylindrical booster with three large stabilising fins.
Guidance and Control: Western Electric Command system. Control by foreplanes.
Warhead: High-explosive, detonated by signal from ground.
Length: 34ft 0in (10.36m).
Body diameter: 1ft 0in (30cm).
Wing span: 4ft 0in (1.22m).
Launch weight: 2,455lb (1,113kg).
Max speed: Mach 2.25.
Max range: 25 miles (40km).

Development and Service
When the first Nike Ajax site became operational in December 1953, it inaugurated America's integrated missile/aircraft air defence system. About 15,000 Nike Ajax missiles were built before the type was superseded in production by the more powerful Nike Hercules; of these 5,500 were fired during development and training. None remain operational in the USA; but Nike Ajax continues in service in a few countries, such as Greece, Italy and Japan, sometimes in a semi-mobile form, although considerable ground equipment is required to support the launchers. The guidance system is similar to that described for Nike Hercules, utilising separate missile and target tracking radars and a computer to work out the necessary command signals to achieve an interception.

Nike Ajax on mobile launcher

Nike Hercules launch-site of the Nationalist Chinese Air Force

Nike Hercules (MIM-14)   USA

Surface-to-air missile. In production and service.

Prime contractors: Western Electric Company Inc/Mitsubishi Jukogyo Kabushiki Kaisha.
Powered by: Thiokol solid-propellant sustainer. Hercules Inc clustered four-motor solid-propellant jettisonable booster.
Airframe: Tandem two-stage missile. Basically-cylindrical missile body, tapering towards nose which carries small cruciform delta surfaces. Long-chord cruciform delta wings, each with trailing-edge control surface. Booster is made up of four cylindrical motors of the type used singly on Nike Ajax, with cruciform stabilising fins indexed in line with missile wings.
Guidance and Control: Western Electric command system. Control by wing trailing-edge surfaces.
Warhead: Alternative nuclear or high-explosive types.
Length: 41ft 6in (12.65m).
Max body diameter: 2ft 7½in (80cm).
Wing span: 6ft 2in (1.88m).
Launch weight: 10,400lb (4,720kg).
Max speed: Mach 3.65.
Max range: 80 miles (130km).

Development and Service

Although development of Nike Hercules (then known as Nike B) was started in 1953, to replace Nike Ajax, it remains in production by Mitsubishi in Japan. In the USA, where it became operational as the nation's primary anti-aircraft defence weapon in 1958, the number of launch-sites has been reduced progressively. In the early 'seventies, only 13 batteries were still active, with a further 27 manned by the National Guard. These will be re-equipped with SAM-D in due course.

The Nike Hercules missile was developed and produced originally by the Douglas Aircraft Company, and achieved a notable early success when one Nike Hercules intercepted another at a range of 30 miles (48km) from its launch-point, and altitude of about 100,000ft (30,500m) over White Sands Missile Range in September 1960.

In operation, the target is acquired first by acquisition radars, which pass it on to the target tracking radar. When the missile is launched, another tracking radar issues command guidance and detonation instructions under the control of the weapon system's data processing equipment. In the USA, Nike Hercules batteries were integrated into the SAGE defence system, which controlled all air defence fighters and missiles. Nike Hercules is also operational in a mobile role, notably with NATO and SEATO forces, using a HIPAR advanced high-power acquisition radar which greatly increases the mobility and detection capability.

Nuclear Falcon (AIM-26)   USA

Air-to-air nuclear armed missile. In service.

Data apply to AIM-26A.
Prime contractor: Hughes Aircraft Company.
Powered by: Thiokol M60 solid-propellant rocket motor.
Airframe: Circular-section body, with maximum diameter at mid-length and tapering towards nose and tail. Rounded nose. Long-chord cruciform delta wings, each with a tail control surface aft of its trailing edge.
Guidance and Control: Hughes semi-active radar homing guidance. Control by tail surfaces.
Warhead: Nuclear, with active proximity fuse.
Length: 7ft 0in (2.13m).
Max body diameter: 11in (28cm).
Wing span: 1ft 8in (0.51m).
Launch weight: 203lb (92kg).
Max speed: Mach 2.
Max range: 5 miles (8km).

Development and Service
The AIM-26A Nuclear Falcon (originally XGAR-11) appeared in 1960, combining the basic control and guidance equipment of the AIM-4A Falcon (page 33) with the vastly-increased destructive power of a nuclear warhead. Radar homing was chosen in preference to infra-red because of its better all-weather capability, longer acquisition range and suitability for attack from any direction, including head-on. Externally, the AIM-26A differed from earlier Falcons in having a more bulbous body, without the nose vanes of AIM-4A/C/D. When the AIM-26A entered service with F-102 Delta Dagger squadrons of USAF Air Defense Command, it was the first nuclear-tipped air-to-air guided missile in the world, just as the AIM-4A had been the USAF's first air-to-air guided weapon of any kind. It was followed into production in 1963 by the AIM-26B (originally GAR-11A) which differs in having a non-nuclear warhead. (See also HM-55, page 34).

Otomat in wind tunnel

Otomat

France/Italy

Ship-to-ship tactical missile. Under development.

Prime contractors: SA Engins Matra (France) and Oto Melara SpA (Italy).
Powered by: Turboméca Arbizon III turbojet, rated at 882lb (400kg) st. Two side-mounted jettisonable boosters.
Airframe: Cylindrical body. Four semi-circular engine air-intake ducts, each supporting a cropped-delta wing, equi-spaced around body mid-way between nose and tail. Cruciform tail control surfaces indexed in line with wings. Ogival nose-cone and tapered tail fairing.
Guidance and Control: Inertial guidance, with Thomson-CSF active radar terminal homing. Tail control surfaces.
Warhead: Semi-armour-piercing type, weighing more than 440lb (200kg).
Length: 15ft 9¾in (4.82m).
Body diameter (forward of ducts): 1ft 3¾in (40cm).
Wing span: 3ft 10¾in (1.19m).
Launch weight: 1,543lb (700kg).
Max speed: Mach 0.82.
Max range: 37-50 miles (60-80km).

Development and Service

This missile takes its name from those of its manufacturers, *Oto* Melara and *Mat*ra. Its development was started in 1969, to provide a weapon in the same category as the French Exocet, but with a much longer range as a result of its use of a turbojet engine instead of a rocket motor. The warhead is unusually large and its effects are increased by the incendiary effect of any fuel remaining in the missile when it strikes the target.

Otomat does not require a swivelling launcher. It can be mounted in a fixed position on the launching craft, inside its delivery container which doubles as launcher. Firing is possible in all weathers, by day or night, and the missile is able to change direction up to 180 degrees, to port or starboard, to align itself on the target, which can therefore be behind the launch-ship. The basic course to the target is worked out by the shipboard radar and fire control system, with radar terminal homing on enemy ships that can be well 'over the horizon' and hence beyond the range of the ship's radar. A radio altimeter maintains a cruising height of 50ft (15m) throughout the missile's flight, with a final attack either from low level or at the end of a steep dive after a climb started 4.5 miles (7km) from the target.

Deliveries will begin in 1973, initially to arm the *Freccia* fast patrol boats of the Italian Navy, each of which will have four launchers.

Penguin launchers on a *Storm* class gunboat

Penguin
Norway

Ship-to-ship tactical missile. In production and service.

Prime contractor: A/S Kongsberg Vaapenfabrikk.
Powered by: Two-stage solid-propellant rocket motor.
Airframe: Cylindrical body with swept cruciform wings mounted mid-way back. Small swept fins on tapered nose, indexed in line with wings.
Guidance and Control: Inertial guidance, with infra-red terminal homing.
Warhead: High-explosive, weighing 264lb (120kg), with contact fuse.
Length: 10ft 0in (3.05m).
Body diameter: 11in (28cm).
Wing span: 4ft 7in (1.40m).
Launch weight: 727lb (330kg).
Max range: 11.5-17 miles (18.5-27km).

Development and Service
This ship-launched tactical missile is already operational on *Storm* class gunboats of the Royal Norwegian Navy, each of which can carry six Penguin launchers on its after deck. The *Snogg* class of torpedo boats are each expected to be fitted with four similar launchers, and others will be carried by five *Oslo* class frigates.

Penguin was developed by the Norwegian Defence Research Establishment, with assistance from the technical branches of the US and West German Navies. Kongsberg Vaapenfabrikk was then entrusted with quantity production, and Penguin is fitted with a warhead similar to that of the Bullpup air-to-surface missile for which Kongsberg was European prime contractor. The container in which it is delivered serves also as the shipboard launcher and embodies the necessary launch-rail. Kongsberg claim that the weapon is compatible with most existing types of naval fire-control systems, and that Penguin is equally suitable for installation on helicopters and other platforms.

Pershing (MGM-31A)
USA

Surface-to-surface, selective-range artillery missile. In service.

Prime contractor: Martin Marietta Corporation.
Powered by: Tandem solid-propellant rocket motors: Thiokol M-105 in first stage and Thiokol M-106 in second stage.
Airframe: Two-stage missile. Basically-cylindrical body, with long tapering pointed nose-cone. Three square vanes at base of second stage, indexed in line with three delta-shape tail control surfaces on first stage.
Guidance and Control: Bendix Eclipse-Pioneer Division inertial guidance system, operating until cut-off and separation of second stage. Control by first-stage tail-fins and jet deflection, and second-stage vanes.
Warhead: Nuclear, in ablative re-entry vehicle.
Length: 34ft 6in (10.51m).
Body diameter: 3ft 4in (101cm).
Launch weight: 10,000lb (4,535kg).
Range limits: 115-460 miles (185-740km).

Development and Service
Martin Marietta's Orlando Division began development of Pershing in 1958, under the technical supervision of the US Army Missile Command, Redstone Arsenal, Alabama. Firing trials started in 1960 and achieved a higher success rate than those for any previous missile, with the result that the weapon was declared operational in 1962, first with the US Army and later, also, with the West German forces.

Pershing is a selective-range weapon, able to provide rapid-reaction support for a theatre or general support for a field army. A typical battalion has four firing batteries, a headquarters battery and a service battery. In its original form, the complete weapon system was mounted on four XM474 tracked vehicles. Under the Pershing 1-A improvement programme, these were replaced by four wheeled vehicles based on the Ford M656 five-ton truck, carrying new support equipment to increase rate of fire and improve system reliability. The Pershing 1-A vehicles comprise the improved erector-launcher, an articulated truck and trailer carrying the missile and its warhead and capable of travel over roads or cross-country; a transporter for the improved programmer/test station and power station; the firing battery control centre truck; and the radio terminal set vehicle with an inflatable antenna. The MGM-31A (originally M-14) operational missile is unchanged by this improvement programme. A training version is in service under the designation MTM-31B (originally M-19).

Pershing 1-A, ready for launch

Phoenix (AIM-54A) USA

Long-range air-to-air missile. In production.

Prime contractor: Hughes Aircraft Company.
Powered by: Rocketdyne Mk 47 Mod 0 solid-propellant rocket motor.
Airframe: Cylindrical body with ogival nose. Long-chord cruciform wings and cruciform tail control surfaces.
Guidance and Control: Hughes radar-homing guidance system. Control by tail surfaces.
Warhead: High-explosive, with proximity fuse.
Length: 13ft 0in (3.96m).
Body diameter: 1ft 3in (38cm).
Wing span: 3ft 0in (0.91m).
Launch weight: 838lb (380kg).
Estimated range: 70-100 miles (110-160km).

Development and Service

Intended originally for the now-abandoned swing-wing F-111B tactical fighter, Phoenix offers such a great increase over the capability of current missiles that it was resurrected as primary armament of the US Navy's Grumman F-14 Tomcat. Maintenance and reliability are improved by the fact that the missile is made up of a number of self-contained sections, so that it can be handled as a complete unit or broken down for easy shipboard check-out and handling. In action, the target is located by the aircraft's Hughes AN/AWG-9 weapon control system, which can lock on to the enemy in any kind of weather and launch Phoenix at optimum range. The pulse-Doppler radar of the AN/AWG-9 enables it to 'look-down' and sort out a moving target from the ground clutter that obscures such targets in other systems. Its track-while-scan mode also enables the fire control system to keep up to six missiles at a time on course while searching for further targets. Terminal homing is by an active radar system in the missile.

Inert Phoenix missiles were first air-launched from an A-3A Skywarrior test aircraft in 1965. By March 1969, it was possible to intercept two Firebee target drones simultaneously with two missiles, while the targets were flying several miles apart. In August 1970, the 'look-down' and track-while-scan features of the AN/AWG-9 were tested in a launch against a Cougar jet fighter/drone flying many miles away and at lower altitude than the F-111B launch aircraft; the Cougar was destroyed. The first Phoenix production contract was placed by the US Navy in December 1970. The missile is expected to become operational on the F-14 in 1973-74.

Pluton

France

Tactical nuclear surface-to-surface missile. In production.

Prime contractor: Aérospatiale, Division Engins Tactiques.
Powered by: SEP Styx dual-thrust solid-propellant rocket motor.
Airframe: Cylindrical body, with pointed ogival nose-cone and tail control surfaces.
Guidance and Control: Simplified inertial guidance system, with SAGEM inertial platform linked to a computer. Control by electrically-actuated tail surfaces.
Warhead: Nuclear, approx 10/15-kiloton yield.
Length: 24ft 10¾in (7.59m).
Body diameter: 2ft 1½in (0.65m).
Fin span: 4ft 8in (1.42m).
Launch weight: 5,335lb (2,420kg).
Max range: 62 miles (100km).

Development and Service

This mobile tactical weapon system is being developed and produced to replace the Honest John bombardment missiles which currently equip five battalions of the French Army. Firing trials of full-scale Pluton test vehicles have been underway since 1969. The versions launched since the Summer of 1970 have been fully representative of the operational missile, complete with guidance system but without a warhead. Meanwhile, development of the nuclear charge has been progressing as part of the French programme of atomic tests in the Pacific, and it is hoped to have Pluton in first-line service by 1973.

Aérospatiale is responsible for the missile and its launching equipment, under the direction of the French Army's Division des Engins Tactiques. The missile and its warhead will be supplied to operational units as two separate packages, with the missile container so designed that it will serve also as the launch-tube when mounted on the tracked transporter-launcher. The latter is based on the AMX-30 tank chassis.

Polaris (UGM-27)

USA

Underwater-to-surface or surface-to-surface ballistic missile. In service.

Data: UGM-27C Polaris A3.
Prime contractor: Lockheed Missiles and Space Company.
Powered by: First stage: Aerojet-General solid-propellant motor. Second stage: Hercules Inc solid-propellant motor.
Airframe: Two-stage missile. Cylindrical body with ogival nose-cone. First- and second-stage casings of glass-fibre. No fins or wings.
Guidance and Control: Inertial guidance system developed under responsibility of Massachusetts Institute of Technology and manufactured by General Electric and Hughes. First-stage control by gimballed nozzles, and second-stage by a thrust-vector system utilising fluid injection.
Warhead: Thermonuclear, in multiple re-entry vehicles (MRV).
Length: 31ft 0in (9.45m).
Max body diameter: 4ft 6in (1.37m).
Launch weight: 35,000lb (15,850kg).
Max speed: 6,600mph (10,620km/h).
Max range: 2,875 miles (4,630km).

Development and Service

When the idea of launching a long-range ballistic missile from a submarine was first suggested, the problems seemed insurmountable. By 1956, advances in solid propellants, guidance systems and warheads offered the possibility of achieving a range comparable with that of land-based intermediate-range ballistic missiles then in production with a weapon only one-quarter their weight. Lockheed Missiles and Space Company (LMSC) began design studies for a missile that would go into a launch-tube on a specially-built submarine. The Autonetics Division of North American and Sperry started developing a Ship's Inertial Navigation System (SINS) that would establish the latitude and longitude of a submarine precisely and so ensure an accurate trajectory for any missile fired from it.

Firings of test missiles ashore and from the surface ship USS *Observation Island,* led to the first underwater launches in 1960. These culminated in the successful firing of two UGM-27A Polaris A1 missiles from a submerged submarine, the USS *George Washington*, on July 20 of that year. This was a nuclear-powered ship which had been modified during construction to have a section containing 16 launch-tubes inserted amidships. The 40 missile submarines which followed for the US Navy, and four for the Royal Navy, were all specially built. The original UGM-27A was eventually superseded by the improved UGM-27B Polaris A2 (weight 30,000lb, 13,600kg; range 1,700 miles, 2,800 km) on 13 submarines, and the UGM-27C (data above) on 28 US and the four British ships.

Polaris A2

Polaris A2 underwater launch

Poseidon (UGM-73)

USA

Underwater-to-surface or surface-to-surface ballistic missile. In production and service.

Data and photograph: UGM-73A Poseidon C3.
Prime Contractor: Lockheed Missiles and Space Company.
Powered by: First stage: Hercules or Thiokol solid-propellant motor. Second stage: Hercules solid-propellant motor.
Airframe: Two-stage missile. Cylindrical body with ogival nose-cone. First- and second-stage casings of glass-fibre. No fins or wings.
Guidance and Control: Inertial guidance system developed under responsibility of Massachusetts Institute of Technology and manufactured by General Electric and Raytheon. Thrust-vector control by one movable nozzle on each stage, actuated by gas generator.
Warhead: Thermonuclear, in multiple independently-targetable re-entry vehicles developed by LMSC and Atomic Energy Commission.
Length: 34ft 0in (10.35m).
Max body diameter: 6ft 2in (1.88m).
Launch weight: approx 65,000lb (12,500kg).
Max range: 2,875 miles (4,630km).

Development and Service

Having established the value of a submarine-launched strategic missile system with Polaris, the US Navy decided in 1965 to evolve the more formidable Poseidon to take advantage of new developments in missile technology. It was possible to pack larger missiles on board existing Polaris A3 submarines with only minor modification to the launch-tubes. This enabled Lockheed Missiles and Space Company (LMSC) to utilise larger motors and improved equipment and so offer double the payload and twice the accuracy of Polaris. When added to the advantages that come from use of an MIRV warhead, these advances make Poseidon far more formidable than its predecessor and it will eventually replace Polaris on 31 of the US Navy's 41 Fleet Ballistic Missile submarines.

Firing trials with Poseidon test rounds began on August 16, 1968. By June 1970 the initial batch of 20 missiles had all been launched, 17 of them from pads ashore and three from tubes on the surface ship USS *Observation Island*. Underwater launch trials began on August 3, 1970, when a Poseidon was fired from the USS *James Madison*, the first submarine that had been fitted with modified launch-tubes and improved navigation and fire control systems for the new weapon. Subsequent development was so rapid that the *James Madison* was declared operational in March 1971.

Quail (ADM-20C)   USA

Air-launched decoy missile. In service.

Prime Contractor: McDonnell Aircraft Corporation.
Powered by: General Electric J85-GE-7 turbojet of 2,450lb (1,112kg) st.
Airframe: Aeroplane configuration. Rectangular body, with rounded nose. Short, high-set, folding cropped-delta wings at rear, each with a vertical tail-fin at mid-span and another projecting downward from the wingtip. Ram air-intake on each side of body.
Guidance and Control: McDonnell autopilot.
Warhead: None, but self-destruct device is fitted to prevent missile being recovered by enemy.
Length: 12ft 10in (3.91m).
Launch weight: 1,200lb (545kg).
Cruising speed: Mach 0.9.
Range: over 250 miles (400km).

Development and Service

Until SCAD is ready for service, Quail will remain quite unique in the missile world. Its sole task is to fly pre-planned 'patterns' in enemy airspace, so that it will be picked up on early warning, missile and interceptor fighter radars and confuse the defences. This is made possible by one of several electronic devices packed into Quail's glass-fibre body, which ensures that the tiny decoy produces the same 'blip' on a radar screen as the huge B-52 Stratofortress bomber from which it is air-launched.

McDonnell began developing Quail, as the XGAM-72, in 1955. By June 1960, a B-52 was able to drop three Quails simultaneously in an operational-type pattern over Eglin AFB. Before the end of that year, the first of the improved production-type ADM-20B Quails showed its paces with a pre-programmed flight of several hundred miles, and deployment to Strategic Air Command bases was completed between March 1961 and January 1962.

Four Quails can be packed into a small launch package, with their wing and tail surfaces folded, for stowage in the bomb-bay of the B-52. Just before launch the engine is started and the wings and fins extend. The current operational version is designated ADM-20C.

R.511 under fuselage of Mirage III-C

R.511   France

Air-to-air missile. In service.

Prime contractor: SA Engins Matra.
Powered by: Hotchkiss-Brandt two-stage solid-propellant rocket motor, with first-stage thrust of 3,530lb (1,600kg) and second-stage thrust of 40lb (200kg).
Airframe: Cylindrical body, with mid-set movable foreplanes on ogival nose. Rear-mounted wings, each with upper and lower fins near tip, and ventral rudder.
Guidance and Control: Semi-active radar homing. Control by foreplanes.
Warhead: High-explosive, with proximity fuse.
Length: 10ft 1½in (3.09m).
Body diameter: 10¼in (26cm).
Wing span: 3ft 3¼in (1.00m).
Launch weight: 397lb (180kg).
Max speed: Mach 1.8.
Max range: 4.5 miles (7km).

Development and Service
Test firings of this first-generation French air-to-air missile began in 1956. It remains in service with Vautour twin-jet all-weather fighter squadrons of the French Air Force, and can be carried by the Mirage III-C, but is obsolescent and limited to pursuit-course attacks. It can be used at heights between 10,000 and 59,000ft (3,000 and 18,000m).

R.530 under fuselage of Mirage

R.530

France

Air-to-air missile. In service.

Prime contractor: SA Engins Matra.
Powered by: Hotchkiss-Brandt two-stage solid-propellant rocket motor of 18,740lb (8,500kg) st.
Airframe: Cylindrical body, with slightly tapering and rounded nose. Cruciform delta wings, of which two have ailerons, indexed in line with cruciform tail control surfaces. Glass-tipped nose on infra-red version.
Guidance and Control: Alternative semi-active radar or infra-red homing. Control via ailerons and movable tail surfaces.
Warhead: Hotchkiss-Brandt, weighing 60lb (27kg), with proximity fuse.
Length: 10ft 9¼in (3.28m).
Body diameter: 10¼in (0.26m).
Wing span: 3ft 7¼in (1.10m).
Launch weight: 430lb (195kg).
Max speed: Mach 2.7.
Max range: 11 miles (18km).

Development and Service

This formidable air-to-air missile is carried as standard armament under the fuselage of Mirage interceptors and under the wings of Vautour IINs of the French Air Force. It has also been exported to nations which have ordered Mirages, including Australia, Israel and South Africa, and is carried on each side of the front fuselage of the French Navy's F-8E(FN) Crusaders. It is an all-weather missile, suitable for use at all heights from sea level to nearly 70,000ft (21,000m). The alternative types of homing head are sensitive enough to pick up the target during attack from any direction; unlike some infra-red missiles which must be launched from behind the target in order to lock on to the heat from its jet efflux.

R.550 under wing of Meteor test aircraft

R.550 Magic   France

Short/medium-range air-to-air 'dogfight' missile. Under development.

Prime contractor: SA Engins Matra.
Powered by: Solid-propellant rocket motor.
Airframe: Slim cylindrical body with rounded nose. 'Double-canard' configuration, with movable foreplane control surfaces indexed in line with cruciform fixed wings and tail-fins.
Guidance and Control: Infra-red guidance system. Control by foreplanes.
Dimensions, Weight and Performance: Secret.

Development and Service
This is Matra's answer to the need for a close-range, fast-manoeuvring 'dogfight' missile, emphasised by the failure of air-launched missiles to hit MiGs during air combats in Vietnam. Its development was started as a private venture in 1967, but has continued under French Air Force sponsorship since 1969. The R.550 can use the same aircraft launcher as the Sidewinder, which it is intended to replace.

Rapier   UK

Surface-to-air close-range missile. In production and service.

Prime contractor: British Aircraft Corporation, Guided Weapons Division.
Powered by: IMI solid-propellant rocket motor.
Airframe: Cylindrical body, with cruciform cropped-delta wings mounted mid-way between the pointed conical nose and the cruciform tail-fins.
Guidance and Control: Radio command guidance system, along line-of-sight to target. Radar tracking system for all-weather operation under development. Control by movable tail surfaces.
Warhead: High-explosive, with impact fuse.
Length: 7ft 3in (2.21m).
Body diameter: 5in (0.13m).
Wing span: 1ft 3in (0.38m).
Max speed: above Mach 2.

Development and Service
Known originally as the ET.316 low-level anti-aircraft guided weapon system, Rapier was developed initially under a contract awarded in September 1964. Progress was so rapid that production of the missile for the British Army and RAF Regiment was able to begin by mid-1967. Operational deployment by both services was under way in 1971, by which time a further large order had been received from the Imperial Iranian Government. Rapier is also reported in service in Zambia.

Rapier is a quick-reaction weapon, designed to achieve a high kill probability when launched against helicopters, subsonic and supersonic fixed-wing aircraft at all heights from ground level to several thousand metres.

The main units of the system comprise a four-round launcher trailer, an optical tracker and a power unit. Once the launcher has been loaded by two members of the five-man detachment, it can be left unattended and the system can be operated by one man at the tracker. When surveillance radar built into the launcher detects a target, an IFF interrogator determines if the latter is hostile. If it is, the launcher and tracker are aligned automatically in the direction of the target. The operator acquires the target optically in elevation and then tracks it by means of a joystick, with the launcher following the tracker as he does so. When the target is within range, as assessed by a computer, the operator is informed and fires a Rapier missile. Flares on the missile are followed by a TV camera in the tracker, which measures deviations from the sightline. The computer uses these data to correct the missile's trajectory and keep it on course to the target. The complete weapon system, with nine additional missiles, can be towed and transported by two Land-Rovers and is air-transportable by two Wessex helicopters.

Rapier fire unit (*left*), tracker in background and generator set (*right*)

Rapier tracker, with fire unit in background

RB04E

RB04   Sweden

Air-to-surface tactical missile. In production and service.

Data apply to RB04E
Prime contractor: Saab-Scania Aktiebolag.
Powered by: Solid-propellant rocket motor.
Airframe: Body of circular section, tapering in a smooth curve towards nose and tail. Mid-set rear-mounted wings, each with a trailing-edge aileron and fixed end-plate fin. Cruciform cropped-delta control surfaces at front of missile.
Guidance and Control: High-efficiency homing guidance. Control by ailerons and foreplanes.
Warhead: High-explosive type, weighing 660lb (300kg).
Length: 14ft 7¼in (4.45m).
Max body diameter: 1ft 7¾in (50cm).
Wing span: 6ft 6in (1.98m).
Launch weight: 1,320lb (600kg).
Performance: Secret.

Development and Service
This formidable all-weather attack missile was evolved originally by the Swedish Guided Weapons Directorate (Robotavdelningen), to meet a 1949 requirement for an anti-shipping weapon. The first complete RB04 was launched from a Saab J029 fighter on February 11, 1955; four years later the missile became standard armament on the Saab A 32A Lansens which equip four attack wings of the Swedish Air Force. Performance is subsonic, but the RB04 has proved so effective that it will continue to form the primary armament of the new Saab AJ 37 Viggen Mach 2 attack fighter, of which production deliveries began in 1971. This aircraft carries a third RB04 under its fuselage, in addition to the usual two on underwing mountings.

Current version of the missile on the A32A is the RB04D, with higher performance than the earlier standard RB04C. The 'D' became operational in 1971 and will be followed by the RB04E, with revised structure and guidance to improve reliability and accuracy. It has been suggested that the 'E' will use the same type of radar homing head as the German Kormoran (page 64).

RB05As under fuselage of Viggen

RB05A
Sweden

Tactical air-to-surface missile. In production.

Prime contractor: Saab-Scania Aktiebolag.
Powered by: Volvo Flygmotor VR-35 storable liquid-propellant rocket motor.
Airframe: Cylindrical body with long ogival nose. Long-chord cruciform delta wings, indexed in line with cruciform tail control surfaces.
Guidance and Control: Command guidance, via pilot-operated joystick, with aid of tracking flare on missile. Control via tail surfaces.
Warhead: High-explosive with proximity fuse.
Length: 11ft 10in (3.60m).
Body diameter: 1ft 0in (0.30m).
Wing span: 2ft 8in (0.80m).
Launch weight: 675lb (305kg).
Performance: Secret.

Development and Service

In its initial form, this weapon lacks the all-weather capability of the RB04, being manually-guided. However, the precision of the guidance system and high degree of manoeuvrability of the missile make it possible to attack sea or land targets situated a considerable distance to either side of the launch aircraft's course, and Saab claims that the RB05A can also be used in an air-to-air role. The microwave radio link over which the pilot's command signals are transmitted is said to be highly resistant to jamming and to permit full control of the missile at low altitudes over all kinds of terrain. The pilot simply guides the missile so that a tracking flare on its rear end remains aligned on the target. Later versions, fitted with a homing guidance system, are being studied to provide for all-weather operation.

Development of the RB05A began in 1960, under the designation Saab 305A. An initial £14.5 million production contract was placed in November 1970, to arm the AJ 37 Viggen and the SK 60B/C versions of the lightweight Saab-105 in Swedish Air Force service.

RB08A

Sweden

Surface-to-surface or ship-to-ship missile. In service.

Prime contractor: Saab-Scania Aktiebolag.
Powered by: Turboméca Marboré IID turbojet Two solid-propellant booster rockets power the launch carriage.
Airframe: Circular-section body which tapers towards nose and tail. Long tapered radome above nose air intake. Mid-set folding wings are swept at 30 degrees and fitted with end-plates and spoilers. Vee-type tailplane and elevators are supplemented by ventral fin, all swept at 30 degrees.
Guidance and Control: Stabilised initially. Terminal homing. Aerodynamic control by spoilers and elevators.
Warhead: High-explosive.
Length: 18ft 9in (5.71m).
Max body diameter: 2ft 2in (0.66m).
Wing span: 9ft 10½in (3.01m).
Launch weight: 1,985lb (900kg).
Performance: Secret.

Development and Service

This missile had its origin in the French CT.20 target drone, designed and manufactured by Nord-Aviation. The French company evolved combat versions known as the SM.20 and MM.20 for surface-to-water (sol-mer) and water-to-water (mer-mer) use respectively, and has stated that these could carry a 550lb (250kg) warhead over a range of 155 miles (250km). The Royal Swedish Navy awarded Saab a contract to develop a similar operational version of the CT.20 in 1959, followed by a production contract in 1965. Deliveries were completed by 1970, and the RB08A is now deployed with coastal defence batteries and on board two destroyers of the Royal Swedish Navy. In the latter case, the missiles are stored below deck and transferred by ramp to the launcher aft of the rear funnel when required for action. The Swedish-designed warhead is claimed to be highly-effective against invasion from the sea.

RB08A on board destroyer

Redeye (MIM-43A) (Swedish Army designation RB69)

USA

Shoulder-fired surface-to-air missile. In service.

Prime contractor: General Dynamics Corporation, Electro Dynamic Division.
Powered by: Atlantic Research Corporation dual-thrust solid-propellant rocket motor.
Airframe: Cylindrical body with flip-out cruciform tail-fins and nose vanes. Glass nose over infra-red seeker.
Guidance: Initial optical aiming, with infra-red terminal homing.
Warhead: High-explosive.
Length (container): 4ft 0in (1.22m).
Body diameter: 2¾in (7cm).
Weight of complete system: 29lb (13.15kg).
Cruising speed: Supersonic.

Development and Service

After satisfying the US Army as to the feasibility of such a weapon system, the former Convair-Pomona division (now Pomona operation of Electro Dynamic Division of GD) received a contract to develop Redeye in August 1959. Production began in 1964 and eventually reached the rate of more than 1,000 missiles per month for the US Army and Marine Corps. Each armoured, artillery and infantry battalion of the Army in Europe has a Redeye section composed of one officer, a sergeant and four to six two-man firing teams. Redeye is also a standard weapon in the Australian and Swedish Armies. Production was completed in the 1970 fiscal year.

Extremely compact, the weapon system consists simply of a sealed launch-tube, containing the missile, and a self-contained optical sighting/firing package attached to the tube. This makes it easily transportable by one man in a combat area, or in rough country, and it is effective against attacking aircraft at altitudes and ranges commensurate with the defence of Army field positions and Marine Corps amphibious operations. When the operator sights an enemy aircraft, he tracks it by means of the sight and energises the missile guidance system. A buzzer informs him when the missile is ready to fire. The boost charge propels the missile out of the launcher. When Redeye has travelled far enough to protect the operator from blast effects, the sustainer fires and propels the missile to the target.

Loading Red Tops on a Sea Vixen

Red Top

UK

Air-to-air missile. In service.

Prime contractor: Hawker Siddeley Dynamics Ltd.
Powered by: Solid-propellant rocket motor.
Airframe: Cylindrical body, with fixed cruciform wings indexed in line with cruciform triangular tail control surfaces. Hemispherical glass nose.
Guidance and Control: Infra-red homing guidance system. Control by tail surfaces.
Warhead: High-explosive, weighing 68lb (31kg).
Length: 11ft 5.7in (3.50m).
Body diameter: 8¾in (22.5cm).
Wing span: 2ft 11¾in (0.91m).
Weight: Secret.
Cruising speed: Mach 3.
Max range: 7 miles (11km).

Development and Service
Red Top was developed originally, as Firestreak Mk IV, to overcome the limitations of the earlier pursuit-course missile. Its much improved infra-red guidance system enables the pilot of the launch aircraft to fire it against the target from any direction, even dead ahead, and the larger wings and tail surfaces offer improved manoeuvrability at all altitudes. Performance is increased by use of a more powerful rocket motor, and the warhead is also heavier than that of Firestreak. Red Top is interchangeable with Firestreak on Lightning F Mk 3 and 6 interceptors of the RAF and Sea Vixen FAW Mk 2s of the Royal Navy.

Roland

France/Germany

Close-range surface-to-air missile. In production.

Prime contractors: Aérospatiale, Division Engins Tactiques/Messerschmitt-Bölkow-Blohm GmbH.
Powered by: Two-stage solid-propellant rocket motor.
Airframe: Slim cylindrical body with pointed ogival nose. Fixed cruciform foreplanes indexed in line with hinged cruciform delta wings which extend as the missile leaves the launch-tube.
Guidance and Control: Radio command guidance by optical aiming/infra-red tracking system. Control by vanes in sustainer motor exhaust.
Warhead: High-explosive, with proximity fuse.
Length: 7ft 10½in (2.40m).
Body diameter: 6.3in (16cm).
Wing span: 1ft 7¾in (0.50m).
Launch weight: 139lb (63kg).
Cruising speed: Mach 1.6.
Range limits: 1,640-19,700 ft (500-6,000m).

Development and Service

Tube-launched from medium-size armoured vehicles, Roland is intended to provide battlefield defence against helicopters and low-flying fixed-wing aircraft flying at speeds up to Mach 1.3. It is one of a family of tactical missiles developed jointly by Aérospatiale and MBB for the French and German Armies, and firing trials have been underway since 1968. In December 1970 four Rolands were launched against a CT.20 target drone; all achieved successful interceptions and the last one, fitted with a warhead, destroyed it. Evaluation was completed in 1971, and current French military budgets provide for the purchase of 65 Roland systems in the period up to 1975.

Roland is fired from a twin-launcher, which can be carried with its pulse-Doppler surveillance and search radar on the French AMX-13 and AMX-30 fighting vehicles and the German SPZ tank. The missile's shipping container serves also as launcher and is loaded by a hydraulic system which recharges the launch-turret each time a missile has been fired. After a target has been located in azimuth by the radar, the operator acquires it in elevation and tracks it optically. His periscopic sight is linked to an infra-red missile tracking system and a computer, which processes the data fed into it and transmits command signals to the missile to keep it on the line-of-sight course to the target.

A radar guidance device, which can be added to the basic system for all-weather capability, has been developed by MBB to meet a specific German requirement. Known as Roland 2, this version is also available as the ship-based Roland 2M.

Safeguard

USA

Anti-ballistic missile defence system. In production.

Prime contractor: Western Electric Company, Inc.

Development and Service

The United States began development of an anti-ballistic missile (ABM) defence system in the mid-1950s to counter the growing capability of the Soviet ICBM force. Known successively as the Nike Zeus, Nike-X, Sentinel and Safeguard, the programme has undergone several stages of major modification, taking into account the crippling cost of attempting to provide an effective defence for anything but America's ICBM sites in the face of Soviet progress with MRV/MIRV, FOBS and other devices. Currently approved (1971-72) programmes provide for full Safeguard protection of two Minuteman ICBM bases at Malmstrom AFB, Montana, and Grand Forks AFB, North Dakota, with a third installation at Whiteman AFB, Missouri, scheduled to follow and advanced site preparation, but no actual construction, authorised near Warren AFB, Wyoming. Each site will deploy a Perimeter Acquisition Radar (PAR), providing initial track data to alert firing units; a Missile Site Radar (MSR) to refine this tracking data and control the flight of ABM missiles to intercept the incoming ICBM re-entry vehicles; and Sprint (page 140) and Spartan (page 139) missiles at selected locations around the bases. Additional Sprints will be installed at four Remote Launch Sites (RLS) located near each MSR. Firing trials of both types of ABM missile have proved successful and the Safeguard installation at Grand Forks is expected to become operational in 1974.

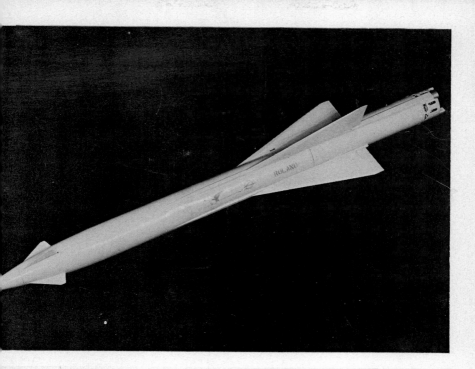

Roland twin-launcher on French AFV

'Sagger on BMP personnel carrier'

Sagger

USSR

Anti-tank missile. In service.

Powered by: Solid-propellant rocket motor.
Airframe: Short cylindrical body, with conical nose. Rear-mounted cruciform wings are swept near body of missile but unswept outboard. Wings fold for storage and transport.
Guidance and Control: Wire guidance.
Warhead: High-explosive.
Length: 2ft 6in (0.76m).

Development and Service
Latest of the three standard types of small wire-guided anti-tank missiles serving with the armed forces of Russia and its allies, 'Sagger' was first displayed publicly in May 1965. It was shown then on a six-round launcher mounted on the BRDM armoured amphibious vehicle which is used also as a carrier for 'Snapper' and 'Swatter'. The installation is interesting, as the launcher is protected during transport by an armoured cover plate that encloses it within the vehicle. When the launcher is elevated for action, the cover plate remains in place above the launcher, affording continued protection. A single 'Sagger' is carried on a rail launcher above the 76mm gun on the BMP eight-man armoured personnel carrier, first seen in 1967 and becoming a standard vehicle throughout the Warsaw Pact nations.

SAM-D

USA

Surface-to-air weapon system. Under development.

Prime contractor: Raytheon Company.
Powered by: Thiokol TX-486 single-stage solid-propellant rocket motor.
Airframe: Cylindrical body, with pointed ogival nose. Cruciform tail control surfaces.
Guidance and Control: Command guidance, with semi-active terminal homing.
Warhead: High-explosive.
Length: 17ft 0in (5.18m).
Body diameter: 1ft 4in (0.41m).

Development and Service
Studies for a new surface-to-air weapon system known as SAM-D began in 1965, following earlier US Army study programmes known as Field Army Ballistic Missile Defense System (FABMDS) and Army Air Defense System for the 1970s (AADS-70). After evaluation of competing industry proposals, Raytheon was awarded in May 1967 a contract covering the first year's advanced development of SAM-D, which is intended to replace Hawk and Nike Hercules. Subsequent contracts have covered demonstration of the advanced guidance system, operation of the radar in its various modes and the aerodynamic capabilities of the missile, leading to 1971 demonstrations of the full capability of the fire control group.

The missile and its shipping/launch container are being developed by Martin Marietta Corporation under sub-contract from Raytheon. Several missiles in their containers will be carried by each SAM-D tracked launch vehicle, from which they will be fired either singly or in close-sequence salvos at selectable azimuths and elevations, as required. In the field, a fire section will be mounted on about four vehicles, including a fire control vehicle carrying radar, radar/weapons control computer, communications and power supply. The multi-function phased-array radar will detect targets, track them, and track and issue guidance commands to the supersonic missile in flight.

Artist's impression of SAM-D battery

Test launch of SAM-D

Samlet   USSR

Surface-to-surface cruise missile. In service.

Powered by: Turbojet engine.
Airframe: Circular-section metal body, with hemispherical radome above chin-type air-intake at nose. Mid-mounted sweptback wings, each with two fences on top surface. Swept horizontal tail surfaces mounted mid-way up tail-fin.
Guidance and Control: Probably beam-riding radar or radio-command guidance with terminal radar homing. Aerodynamic control surfaces.
Warhead: High-explosive.
Length: 27ft 0in (8.2m).
Wing span: 16ft 0in (4.9m).
Max range: 63 miles (100km).

Development and Service
This subsonic cruise missile is a surface-to-surface version of the air-to-surface 'Kennel' (see page 62). It differs little from the air-launched weapon and is used primarily for coastal defence by the armed forces of the Soviet Union, Poland and Cuba.

Sandal (SS-4) — USSR

Surface-to-surface intermediate-range ballistic missile. In service.

Powered by: One liquid-propellant engine. No booster.
Airframe: Cylindrical metal body, with flared skirt around nozzle and small cruciform tail-fins. Pointed conical nose-cone.
Guidance and Control: Probably radio-inertial guidance system. Control by elevators in trailing-edges of tail-fins and vanes operating in rocket efflux.
Warhead: Alternative nuclear or high-explosive.
Length: 68ft 0in (20.8m).
Max body diameter: 5ft 3in (1.60m).
Launch weight: about 60,000lb (27,200kg).
Max speed: 4,300mph (6,900km/h).
Max range: 1,100 miles (1,750km).

Development and Service

After early post-war firing trials with captured German V-2 bombardment rockets, the Soviet Union evolved an improved version which became known to NATO forces as the SS-3 'Shyster'. This had a longer, cylindrical body, permitting increased propellant tankage, and was first displayed in a Moscow parade in November 1957. Its transporter was hauled by a tracked vehicle carrying the launch crew.

In 1961, a further-improved version was shown for the first time. Known in the west as the SS-4 'Sandal', this is longer, giving increased propellant capacity and range. It can be identified by the flared skirt over its rocket nozzle, and smaller tail-fins. 'Sandal's' transporter is hauled by the same kind of vehicle as that of 'Shyster', the complete missile system being completed by other tractor/trailer units carrying ground support equipment and propellants.

'Sandal' replaced 'Shyster' as Russia's standard IRBM and has been used widely, in modified forms, as a satellite launch vehicle. About 600 remain operational, of which some ten per cent are directed at China and Japan, and the remainder at targets in western Europe. Discovery by US reconnaissance aircraft of an attempted deployment of 'Sandals' in Cuba led to the Soviet/American confrontation in 1962 and eventual withdrawal of the missiles.

Sark (SS-N-4)　　　　　　　　　　　　　　　　　USSR

Underwater-to-surface or surface-to-surface missile. In service.

Powered by: Solid-propellant motors, with seven cylindrical first-stage nozzles.
Airframe: Probably two-stage missile. Basically cylindrical body, with second stage of slightly smaller diameter. Blunt conical nose-cone. No fins or wings.
Warhead: Nuclear.
Length: 48ft 0in (14.50m).
Max body diameter: 5ft 9in (1.75m).
Weight and Performance: Unknown.

Development and Service
This first-generation Soviet counterpart of Polaris was first included in a Moscow parade on November 7, 1962. The commentator said that it could be fired either underwater or on the surface. Projections on each side of the casing are assumed to be used to locate the missile inside its launch-tube.

Before the Soviet-Chinese diplomatic split, a single 'G' class submarine, with three missile launch-tubes, was supplied by Russia to the Chinese Navy. The missiles carried by this ship are said to have a range of about 380 miles (600km) and may well be 'Sarks'.

Sasin (SS-8)　　　　　　　　　　　　　　　　　USSR

Exhibited as intercontinental ballistic missile. Not thought to be an operational weapon.

Powered by: Storable liquid-propellant rocket motors.
Airframe: Two-stage rocket. Both stages are cylindrical, with tapered fairing between first stage and smaller-diameter second stage. Blunted conical nose cone. No skirt, fins or wings.
Warhead: Thermonuclear.
Length: 80ft 0in (24.40m).
Body diameter (first stage): 9ft 0in (2.75m).
Max range: 6,500 miles (10,500km).

Development and Service
Although this rocket was exhibited in several Moscow parades, from November 1964 onwards, it has not been seen recently and there is little to suggest that it ever became an operational weapon. On the other hand it has a Western missile designation (SS-8) and the International Institute for Strategic Studies continues to list a total of 220 ICBMs of the SS-8 and similar SS-7 'Saddler' types in the current Soviet long-range missile force. If this were correct, these weapons could be expected to carry a thermonuclear warhead of about 5-megaton yield.

Savage (SS-13)

USSR

Intercontinental ballistic missile. In production and service.

Powered by: Three solid-propellant stages, each with four nozzles.
Airframe: Three-stage missile, with truss structure between stages. Each stage is cylindrical, of progressively-reduced diameter, with a flared skirt over its nozzles. Pointed conical nose-cone.
Warhead: Thermonuclear or nuclear.
Length: 66ft 0in (20.0m).
Body diameter (excluding skirt):
1st stage 5ft 6in (1.68m).
2nd stage 4ft 7½in (1.40m).
3rd stage 3ft 2½in (0.98m).
Max range: 5,000 miles (8,000km).

Development and Service
This Russian counterpart of Minuteman was first displayed in a Moscow military parade celebrating the 20th anniversary of the end of the second World War in Europe, on May 9, 1965. It is intended to replace the storable liquid-propellant SS-11 and up to 100 were thought to be operational in 1971. Firing trials down the Pacific range, from Tyuratam in Kazakhstan, are continuing. Two SS-13s tracked by America in November 1970 each delivered a single inert warhead over a range of about 4,500 miles (7,250km).

The top two stages of 'Savage' are used without the first stage in the 'Scamp/Scapegoat' mobile missile system.

Sawfly (SS-N-6) USSR

Underwater-to-surface or surface-to-surface missile. In service.

Powered by: Solid-propellant motors, with four first-stage nozzles.
Airframe: Probably two-stage missile. Cylindrical body of constant diameter and blunt, tapered nose-cone. No fins or wings.
Warhead: Nuclear or thermonuclear.
Length: 42ft 0in (12.8m).
Max body diameter: 5ft 9in (1.75m).
Range: At least 1,725 miles (2,780km).

Development and Service

None of the three types of submarine-launched ballistic missiles included in Moscow parades since 1962 is believed to be truly representative of an operational weapon. Thus, although 'Sawfly' may give a general idea of the appearance of the most formidable of the trio, its nose-cone is almost certainly an inert test package.

'Sawfly' is considerably larger than the US Navy's Poseidon; but it is estimated to have a shorter range because American propellants are still generally superior to those used in Soviet rockets. However, its availability is providing the Soviet Navy with a formidable attack capability. There were estimated to be eighteen 'Y' class nuclear-powered submarines in service by early 1971, each armed with 16 'Sawflies'. Anything from 35 to 50 such vessels could be in service by the mid-seventies.

SCAD
USA

Subsonic cruise armed decoy missile. Under development

Powered by: Turbojet engine.
Guidance and Control: Inertial guidance system, with terminal assistance.
Warhead: Nuclear.
Design range: 975 miles (1,575km).

Development and Service
This projected USAF missile was inspired by the success of Quail (page 94). The basic subsonic cruise armed decoy (SCAD) version is intended initially as part of the ECM equipment carried by B-52 Stratofortress strategic bombers, and later for carriage by the swing-wing supersonic North American Rockwell B-1. An unarmed (SCUD) version has also been proposed.

Scaleboard (SS-12)
USSR

Surface-to-surface tactical missile. In service.

Powered by: Storable liquid-propellant rocket motor.
Airframe: Cylindrical body, with tapered nose-cone and either tail-fins or vernier engines at base.
Guidance and Control: Probably inertial guidance, set up before missile launch by means of equipment situated between two inner wheels on port side of transporter.
Warhead: Probably thermonuclear.
Length: 37ft 0in (11.25m).
Max range: 450 miles (725km).

Development and Service
First seen in November 1967, this mobile ballistic missile is about the same length as 'Scud-B' (page 120) and is transported on a similar MAZ-543 eight-wheeled erector-launcher. It is, however, of larger diameter than the 'Scuds', implying a longer range, and differs in being enclosed in a ribbed casing which is elevated with it into the firing position. No photographs have yet shown the missile without this casing. During transit, the driver of the vehicle occupies the left-hand cab at the front, with the launch operator and his control console in the right-hand cab and seats for at least three other members of the launch crew to the rear of these cabs. An increasing proportion of the total of 300 short-range ballistic missiles estimated to be in service with Soviet land forces is likely to consist of 'Scaleboards'.

Scamp/Scapegoat (SS-14)   USSR

Mobile surface-to-surface ballistic missile system. In service.

The following data refer to the 'Scapegoat' missile.
Powered by: Solid-propellant motors, with four nozzles on each stage.
Airframe: Two-stage missile, with truss structure between stages. Each stage is cylindrical with a flared skirt over its nozzles. Second stage has smaller diameter than first, and is fitted with a pointed conical nose-cone.
Warhead: Thermonuclear or nuclear.
Length: 35ft 0in (10.6m).
Body diameter (excluding skirt):
1st stage 4ft 7½in (1.40m).
2nd stage 3ft 2½in (0.98m).
Max range: 2,500 miles (4,000km).

Development and Service

First included in a Moscow parade to mark the 20th anniversary of the end of the second World War, on May 9, 1965, this tracked weapon system was nicknamed 'Iron Maiden' by Western observers. This reflected the fact that the missile is enclosed in a container which is made in two halves, split horizontally and hinged. For firing, a massive hydraulic jack on each side of the vehicle, at the rear, raises the container to a vertical position. The container is then opened and moved away from the missile before the latter is launched.

For several years the configuration of the missile carried by the tracked erector/launcher was not known; the complete weapon system was therefore given the NATO code-name 'Scamp'. The separate code-name 'Scapegoat' was given to a two-stage IRBM first displayed on a trailer in 1967, which appeared to consist of the top two stages of the ICBM known as 'Savage' (see page 111). Not until later did official Soviet films of missile exercises reveal 'Scapegoat' to be

the missile carried inside the 'Scamp' container. The original type of tracked vehicle has been superseded by an improved model, similar to that used in the 'Scrooge' weapon system. Like the latter, 'Scamp/Scapegoat' has been reported in operational service near Buir Nor in Outer Mongolia, adjacent to the Chinese frontier.

Scarp (SS-9/FOBS) USSR

Intercontinental ballistic missile, and fractional orbital bombardment system. In production and service.

Powered by: Liquid-propellant rocket motors. Ring of six first-stage nozzles is surrounded by four equi-spaced vernier nozzles.
Airframe: Cylindrical body of constant diameter, which tapers near the top to blend with cylindrical re-entry vehicle of much smaller diameter. Wedge-shaped fairings over vernier nozzles at tail. No fins or wings.
Warhead: At least three alternative warheads have been tested: a single 20/25-megaton charge; an unguided MRV comprising three separate re-entry vehicles, each with a yield of 5 megatons; and the FOBS 'space-bomb'. An MIRV warhead is under development.
Length: 113ft 6in (34.5m).
Body diameter: 10ft 0in (3.05m).
Max range: approx 5,500 miles (8,800km).

Development and Service

'Scarp' is the most formidable weapon that has yet been revealed anywhere in the world, with a thermonuclear warhead of 20/25-megaton yield. It was first exhibited in the Moscow parade of November 7, 1967, marking the 50th anniversary of the Communist Revolution. Four days earlier, the American Secretary of Defense had announced the existence of a Soviet 'space bomb' which was first tested, in the guise of the Cosmos 139 satellite, on January 25, 1967. During many subsequent tests, the inert 'space bomb' payloads have been sighted optically from the Royal Aircraft Establishment at Farnborough, and are estimated to be about 6ft 6in (2.0m) long and 4ft (1.2m) in diameter. The 'space bomb' version of 'Scarp' is known in America as a fractional orbital bombardment system (FOBS). This reflects the fact that the rocket puts its thermonuclear payload into an orbit about 100 miles (160km) above the Earth. At a pre-determined point, before completion of the first orbit, the payload is intended to be slowed by retro-rocket so that it will drop precisely on to its target.

As an alternative to its single very large warhead, the standard SS-9 ICBM can carry a simple, unguided multiple re-entry vehicle (MRV) made up of three separate smaller charges. During one test in 1970, observed by US reconnaissance satellites and naval vessels, an MRV travelled about 5,500 miles (8,800km) and re-entered the atmosphere with its three inert warheads dispersing to impact 80 miles (130km) apart—about the same distance that separates a cluster of three typical US Minuteman ICBM silo launchers. There is also reason to believe that 'Scarp' has been used as launcher for the Cosmos satellites which have intercepted and destroyed other test satellites in orbit. About one-quarter of the 1,200 Soviet ICBMs in service in late 1970 were SS-9s.

Scrag USSR

Exhibited as intercontinental ballistic missile. Not thought to be an operational weapon.

Powered by: Liquid-propellant rocket motors. Four nozzles on first stage, one large nozzle on second stage, one smaller nozzle on third stage.
Airframe: Three-stage missile, with truss structure between stages. Each stage is cylindrical, the first two of constant diameter, with flared skirt around first-stage nozzles only. Pointed ogival re-entry vehicle on nose of short third stage. No fins or wings.
Guidance and Control: First-stage control by four gimballed nozzles.
Length: 120ft 0in (36.5m).
Body diameter:
1st stage (over skirt) 9ft 5in (2.85m).
2nd stage 8ft 9in (2.70m).
3rd stage 7ft 9in (2.35m).
Max range: 5,000 miles (8,000km).

Development and Service
When 'Scrag' was first shown in a Moscow parade, on May 9, 1965, it was described as a sister vehicle to the launcher used for Yuri Gagarin's Vostok spacecraft. It bears a superficial resemblance to the vehicles used also to launch many other Soviet scientific spacecraft, including Voskhod, Soyuz and the Venus probes. There is, however, no evidence that 'Scrag' has been put into production as an ICBM, or has even been test-fired as such, despite Soviet representation of it as a weapon. The re-entry vehicle would be large enough to accommodate a very powerful nuclear or thermonuclear charge if this rocket had been put to military use.

Scrooge

Mobile strategic weapon system. In service.

Warhead: Thermonuclear or nuclear.
Length of launch-tube: 62ft 0in (18.90m).
Diameter of launch-tube: 6ft 6in (2.00m).
Max range: 3,500 miles (5,600km).

Development and Service

Despite an obvious similarity in concept to the 'Scamp/Scapegoat' weapon system, 'Scrooge' remains one of the Soviet Union's lesser-known strategic missiles. The container, housing the missile, is raised vertically for firing, but does not hinge open and is used as a launch-tube. Its length suggests that the missile is much longer than 'Scapegoat' and would therefore have a greater range. Otherwise, little is known about 'Scrooge' except that it has been observed in first-line service near Buir Nor, on the frontier between Outer Mongolia and Communist China.

Scrubber (SS-N-1)

Ship-based cruise missile. In service.

Airframe: Aeroplane configuration; larger than 'Styx'.
Warhead: High-explosive.
Length: 24ft 11in (7.6m).
Max range: 150 miles (240km).

Development and Service

Although 'Scrubber' was one of the first missiles deployed on ships of the Soviet Navy, very little is known about it. A *Guide to the Soviet Navy*, published by the US Naval Institute, Annapolis, suggests that it is based on the wartime German V-1 flying bomb and the length and range quoted above are taken from this report. 'Scrubber' is housed in a large container/launcher, with projecting ramp, on the aft deck of the four *Kildin* class ships. The five destroyers of the *Krupny* class each have two similar launchers, fore and aft.

Scud-A

Scud-A (SS-1b) — USSR

Surface-to-surface missile. In service.

Powered by: Storable liquid-propellant rocket motor.
Airframe: Cylindrical single-stage metal structure, with movable cruciform tail-fins and pointed nose-cone.
Guidance and Control: Reported to utilise command guidance system, possibly superseded by simplified inertial system. Control via tail surfaces.
Warhead: Nuclear, possibly with alternative high-explosive type.
Length: 35ft 0in (10.66m).
Body diameter: 2ft 6in (0.75m).
Launch weight: 10,000lb (4,500kg).
Max speed: Mach 5.
Max range: 80 miles (130km).

Development and Service
This heavy artillery rocket was first revealed in public in 1957, on the same occasion as 'Frog-1' and 'Frog-2'. It utilises the same basic launch vehicle as 'Frog-1', but differs from all the members of that family by embodying some form of guidance system. The tubular-metal cradle on which 'Scud-A' is supported during transportation is an integral part of the erection gear and serves as a ladder to give access to the warhead when the missile is elevated prior to launch. Preparation for firing is said to take more than one hour after the weapon system reaches the launch-site.

'Scud-A' serves with several Warsaw Pact nations other than the Soviet Union, possibly with high-explosive warhead.

Scud-B (SS-1c)

USSR

Surface-to-surface missile. In service.

Powered by: Storable liquid-propellant rocket motor.
Airframe: Cylindrical single-stage metal structure, with movable cruciform tail-fins and pointed nose-cone.
Guidance and Control: Probably a simplified inertial guidance system. Control via tail surfaces.
Warhead: Nuclear.
Length: 37ft 0in (11.25m).

Development and Service
Enlarged to provide increased propellant tankage, and hence longer range, this improved 'Scud' first appeared in November 1965, carried on the then-new MAZ-543 eight-wheeled transporter/erector/launch vehicle. It is supported on a similar cradle to that used for 'Scud-A', with integral ladders at the sides.

Seacat (Swedish Navy designation RB07) — UK

Surface-to-air and surface-to-surface ship-based close-range missile. In production and service.

Prime contractor: Short Brothers and Harland Ltd.
Powered by: IMI two-stage solid-propellant rocket motor.
Airframe: Basically-cylindrical body, changing to wider and more square cross-section on forward portion and with a pointed ogival nose. Pivoted cruciform sweptback wings, indexed at 45 degrees to fixed cruciform tail-fins, two of which carry tip-mounted tracking flares.
Guidance and Control: Radio command, with visual or radar tracking. Control by pivoted wings.
Warhead: High-explosive, with contact and proximity fuses.
Length: 4ft 10.3in (1.48m).
Max body diameter: 7.5in (19.05cm).
Wing span: 2ft 1.6in (0.64m).
Weight: Approx 150lb (68kg).
Max range (estimated): 2.2 miles (3.5km).

Development and Service

Seacat has proved one of the most widely-exported guided weapons in the world. Its development started under a contract awarded in April 1958. Within four years sea trials were underway from HMS *Decoy*. Today, in addition to the Royal Navy, a total of 14 other nations deploy this small and highly-manoeuvrable close-range weapon system on their naval craft. Production continues and Seacat is expected to remain in service well into the 'eighties.

Simplest of many different fire control systems in service is the Mk 20, in which the standard four-round launcher is controlled by two men in a director unit. One rotates the director so that the second (the aimer) can pick up and track the target through binoculars. The launcher is 'slaved' to the director in azimuth and elevation, with the result that missiles enter the aimer's field of vision soon after launch. The aimer guides each missile to the target by means of a small joystick, using flares on the missile as a tracking reference. The Mks 21 and 22 radar systems provide auto-follow of the target, permitting both visual and 'dark' firings. Sweden and Chile use a Dutch M4/3 radar system; Argentina uses an Italian system. Bofors of Sweden have developed a combined gun/missile system for Germany, with Seacat providing the missile element. Brazil and Iran utilise a lightweight Seacat system, with three-round launcher, which can be accommodated on vessels as small as fast patrol boats and minesweepers. In addition Seacat missiles are used in the Tigercat and Hellcat weapon systems.

Seacat launch from the New Zealand frigate HMNZS *Taranaki*

Sea Dart (CF.299)

UK

Ship-based surface-to-air and surface-to-surface missile. In production.

Prime contractor: Hawker Siddeley Dynamics Ltd.
Powered by: Rolls-Royce Odin ramjet and tandem jettisonable IMI solid-propellant rocket booster. Between-stage vents enable the Odin to be started prior to staging.
Airframe: Missile has cylindrical body with four interferometer aerials around nose intake for ramjet. Rear-mounted long-chord cruciform wings indexed in line with cruciform tail control surfaces. Larger-diameter cylindrical booster has cruciform fins that fold forward while the missile is on its launcher.
Guidance and Control: Marconi semi-active radar homing guidance. Control by tail surfaces.
Warhead: High-explosive, with EMI proximity fuse.
Length: 14ft 3½in (4.36m).
Body diameter: 1ft 4½in (42cm).
Launch weight: 1,210lb (550kg).
Range: At least 19 miles (30km).

Development and Service

Development of Sea Dart (originally CF.299) was started in 1962, to provide a weapon that would represent a major improvement on Seaslug and yet be compact enough to install on ships smaller than those fitted with the earlier missile. Choice of a ramjet sustainer has made Sea Dart a genuine area-defence weapon, capable of intercepting aircraft and air-launched or surface-launched missiles at both very high and extremely low altitudes, with added effectiveness against surface targets. Firing trials have been underway since 1965, confirming that the weapon system offers a rapid launch-rate and the capability of dealing with many targets simultaneously. Production was initiated in 1967 and the first ship to be fitted with the Sea Dart system is the guided missile destroyer HMS *Bristol*, which has a twin-launcher. A similar launcher is being fitted on each of the four destroyers of the new *Sheffield* class, and on two further ships

of this class for the Argentine Navy. The Sea Dart system utilises a surveillance radar, two Marconi Type 909 tracking/illuminating radars, a Vickers twin-launcher and handling system, a Ferranti computer for target selection, data handling, missile launch and in-flight control, and Plessey Radar operations room display equipment.

Sea Indigo

Italy

Ship-based surface-to-air short-range missile. Under development.

Prime contractor: Sistel—Sistemi Elettronici SpA.

Development and Service
Sea Indigo is a navalised version of Indigo (see page 59), with modifications to suit it for the change of role. It can be utilised in launching systems with either automatic or manual reloading, with the latter recommended for ships of under 500 tons displacement. Indigo's Super Fledermaus fire control system is replaced by the Sea Hunter type specified for Sistel's Sea Killer missile systems.

Sea Killer Mk 1

Italy

Short-range surface-to-surface ship-launched missile. In service.

Prime contractor: Sistel—Sistemi Elettronici SpA.
Powered by: Solid-propellant rocket motor, rated at 4,410lb (2,000kg) st.
Airframe: Cylindrical light alloy body with ogival nose-cone. Pivoted cruciform wings mounted mid-way back on body and indexed in line with cruciform tail-fins.
Guidance and Control: Guidance by beam-riding/radio command, in conjunction with radar altimeter. Control by movable wings.
Warhead: High-explosive fragmentation type, detonated by impact/proximity fuse.
Length: 12ft 3in (3.73m).
Body diameter: 7.87 in (0.20m).
Wing span: 2ft 9½in (0.85m).
Launch weight: 370lb (168kg).
Max speed: Mach 1.9.
Max range: 6.2 miles (10km).

Development and Service
Known for a time as Nettuno, this missile is carried in a five-round launcher on board the Italian Navy's fast patrol boat *Saetta*. It is a fully developed operational weapon and can be integrated with existing naval fire control systems such as the Contraves Italiana Sea Hunter 2 and the Contraves AG (Switzerland) Sea Hunter 4.

Sea Killer Mk 1 launcher on the fast patrol boat *Saetta*

Sea Killer Mk 2

Italy

Surface-to-surface ship-launched missile. Available for service.

Prime contractor: Sistel—Sistemi Elettronici SpA.
Powered by: SEP 299 solid-propellant booster, rated at 9,702lb (4,400kg) st for 1.6 seconds, and SEP 300 solid-propellant sustainer, rated at 220lb (100kg) st for 73 seconds.
Airframe: Cylindrical light alloy body, with ogival nose-cone. Pivoted cruciform wings of rectangular shape mounted mid-way back on second stage and indexed in line with cruciform tail-fins. Tandem booster with large rectangular cruciform fins indexed in line with second-stage surfaces.
Guidance and Control: Guidance by beam-riding/radio command, in conjunction with radar altimeter. Control by movable wings.
Warhead: High-explosive, with impact/proximity fuse. Designed to penetrate steel armour.
Length: 14ft 9in (4.50m).
Body diameter: 7.87in (0.20m).
Wing span: 2ft 9½in (0.85m).
Launch weight: 530lb (240kg).
Max range: 11.5 miles (18.5km).

Development and Service
This Mk 2 version of Sea Killer is a two stage missile, with tandem booster, enabling it to carry a heavier warhead and giving a greatly increased range, at a high subsonic cruising speed. Flight trials have been under way since mid-1969, followed by the first weapon system qualification trials in mid 1971. Sea Killer Mk 2 can be integrated with the same Contraves shipboard systems as the Mk 1 version.

Seaslug

UK

Surface-to-air and surface-to-surface ship-launched missile. In service.

Prime contractor: Hawker Siddeley Dynamics Ltd.
Powered by: ICI solid-propellant rocket motor. Four jettisonable solid-propellant wrap-round boosters.
Airframe: Cylindrical light alloy body with pointed nose-cone. Cruciform wings mid-way back on body, indexed in line with cruciform pivoted tail control surfaces. Flat-nose boosters extend forward from wings to nose of missile.
Guidance and Control: General Electric beam-riding guidance system, utilising Type 901M shipboard radar. Control via movable tail surfaces.
Warhead: High-explosive, with proximity fuse.
Length: 20ft 0in (6.10m).
Body diameter: 1ft 4.1in (0.41m).
Wing span: 4ft 8.6in (1.44m).
Weight and performance: Secret.

Development and Service

Standard long-range anti-aircraft missile armament in the Royal Navy, Seaslug is installed on twin launchers in each of the eight fleet escort super-destroyers of the *County* class. The first four of these ships have the Mk 1 version of Seaslug, which achieved a 90 per cent success rate in launchings against fast target drones at both high and low altitudes during its development testing. The Seaslug Mk 2, in the last four ships, offers longer range, Improved performance against low-level targets, and surface-to-surface capability.

Sea Sparrow (RIM-7H)

USA

Ship-launched surface-to-air close-range missile. In production and service.

Prime contractor: Raytheon Company.

Development and Service

Work on Sea Sparrow began in 1964, under the US Navy designation Basic Point Defence Missile System (BPDMS), to provide protection against enemy aircraft and cruise missiles for ships lacking other missile systems. To save time, the Basic version utilises standard Sparrow III missiles (see page 138) in a modified eight-round Asroc launcher (see page 15). The complete system is mounted on a modified automatic gun carriage and weighs 39,000lb (17,690kg). The Sparrow's CW semi-active radar homing system is unchanged. Shipboard equipment includes a Mk 51 director/illuminator to acquire and illuminate the target, the manually-operated Mk 115 fire control system being supplied with the necessary target data from the ship's combat information centre. First ships to be fitted with the BPDMS were US Navy attack carriers. Altogether, 75 ships will eventually carry the system.

An improved PDMS, known also as NATO Sea Sparrow, is being developed by Raytheon in partnership with companies in Denmark, Italy, the Netherlands and Norway under an agreement signed in 1968. This retains the Sparrow missile in conjunction with a new lightweight eight-round launcher, a fire control system based on digital computers, a target acquisition system and a powered director/illuminator. Manufacture was expected to begin in 1972.

Independently of the IPDMS, the USA is developing an Advanced PDMS which will utilise a new missile of similar size to Sparrow. This is unlikely to be available for operational use before the late '70s.

Sea Sparrow on the amphibious assault ship USS *Okinawa*

Seawolf

UK

Close-range surface-to-air and surface-to-surface missile. Under development.

Prime contractor: British Aircraft Corporation, Guided Weapons Division.

Development and Service
BAC received a contract in June 1967 to develop an advanced surface-to-air weapon system which was then designated PX430 and is now known as Seawolf. Intended as a follow-on to Seacat (page 121) it will be an all-weather system, giving greatly improved defence capability against supersonic anti-shipping missiles and aircraft, and will be suitable for surface-to-surface use against ships and hovercraft. Marconi have responsibility for the entire electronics system, including surveillance radars, target tracking radar, missile gathering and guidance TV and data handling equipment.

Few details of the Seawolf missile are yet available, although development is at an advanced stage. It has been stated officially that the missile embodies techniques already used successfully in Rapier (see page 97) and that once a target has been identified as hostile all subsequent phases of launch and guidance are automatic. The Vickers-designed Mk 25 Mod O launcher has six rectangular launch-tubes.

Model of Sea Wolf

Serb (SS-N-5)   USSR

Underwater-to-surface or surface-to-surface missile. In service.

Powered by: Solid-propellant motors. Eighteen small gas-jets at base for ejection from launch-tube.
Airframe: Two-stage missile. Both stages and round-nosed re-entry vehicle are cylindrical, of progressively-reduced diameter, with tapered fairing between each. No fins or wings.
Warhead: Nuclear or thermonuclear.
Length: 35ft 1in (10.7m).
Max body diameter: 5ft 0in (1.50m).
Max range: 750 miles (1,200km).

Development and Service
This Polaris-type weapon was first shown in a Moscow parade in November 1964, suggesting that it represents an intermediate stage between the SS-N-4 'Sark' and the SS-N-6 'Sawfly'. It was described from the start as being suitable for underwater launch from Soviet submarines. After three years, in a 1967 parade, the usual cover-plate over the base was removed to reveal the nozzles for the electrically-detonated cold-gas system by which 'Serb' is ejected from its launch-tube for above-the-surface ignition of the first-stage motor.

Like 'Sark', this missile can be carried by Soviet 'G II' class submarines, in two or three vertical launch-tubes, and by 'G III' and 'H II' class ships, each of which has three tubes.

Sergeant (MGM-29A)   USA

Medium-range surface-to-surface field artillery missile. In service.

Prime contractor: Sperry Rand Corporation, Univac Salt Lake City.
Powered by: Thiokol M-100 solid-propellant rocket motor, rated at 45,000lb (20,400kg) st.
Airframe: Cylindrical steel body. Cruciform tail-fins have small control surfaces hinged to trailing-edges and linked to jet-deflection vanes operating in rocket efflux. Pointed nose-cone.
Guidance and Control: Inertial guidance system made by Univac Salt Lake City. Control by hinged tail surfaces and jet-deflection vanes.
Warhead: Alternative nuclear or high-explosive.
Length: 34ft 6in (10.51m).
Body diameter: 2ft 7in (0.79m).
Fin span: 5ft 10.2in (1.78m).
Launch weight: 10,100lb (4,580kg).
Range limits: 28-85 miles (46-135km).

Development and Service

Sergeant was developed as a more mobile, quicker-reaction replacement for the earlier liquid-propellant Corporal artillery missile. Deliveries began in 1961 and it became operational in the following year. About 500 are thought to equip US Army units, with a further 100 in service with the West German Army. They are air-transportable, and each can be emplaced and fired in a few minutes by a six-man crew.

In recent years, since production ended, Univac Salt Lake City have concentrated on improving the efficiency of the weapon system. Faster count-down and firing rate resulted from introduction of a new digital computer, which replaced several of the original items of ground equipment. By integrating the functions of other electronic equipment, it was possible to dispense with two vehicles, so that each Sergeant battery now needs only one 2½-ton truck and three semi-trailers.

Shaddock (SS-N-3)

USSR

Surface-to-surface cruise missile. In service.

Powered by: Ramjet or turbojet cruise motor, plus two large solid-propellant JATO boosters.
Airframe: Basically bullet-shape body, probably with two short-span hinged wings at mid-point and with pointed ogival nose-cone. Two jettisonable boosters under tail, to each side of large ventral fin.
Guidance and Control: Active radar homing system.
Warhead: Probably nuclear.
Length: 35ft 9in (10.9m)
Max speed: Mach 0.85.
Max range: 230 miles (370km).

Development and Service

Cruise missiles of the subsonic 'flying bomb' type continue to serve in important roles in the Soviet armed forces. None is more formidable than 'Shaddock', which is carried as a standard weapon by many surface ships and submarines. Latest available non-classified data suggest that the four large destroyers known to NATO as the *Kresta I* class each carry four 'Shaddocks' in two twin launch-tubes, without reload capability. This armament is more than doubled on the four destroyers of the *Kynda* class, each of which has four launch-tubes in a side-by-side cluster on the foredeck and four more on the after deck, almost certainly with eight more missiles in store for a second strike. At least 59 submarines are thought to be armed with 'Shaddock'. The nuclear-powered 'E I' and 'E II' classes carry six and eight missiles respectively; the 'J' class carry four each, and the converted 'W' class each carry two or four in what are known as longbin containers. About 100 more 'Shaddocks' are believed to be land-based.

'Shaddock' has never been revealed in public and the above description is based on what can be seen through the open ends of missile containers mounted on trucks which have taken part in parades through Moscow. The drawing is based on one which appeared in an official East German magazine, but has been modified to conform with known details of the nose and tail configuration. No details of the air intake for the cruise engine are known, but this could be annular like that of 'Ganef' (see page 43).

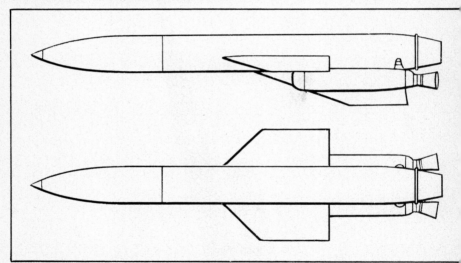

Shillelagh (MGM-51A) USA

Lightweight army close-support missile. In service.

Prime contractor: Philco-Ford Corporation, Aeronutronic Division.
Powered by: Amoco Chemicals single-stage solid-propellant rocket motor.
Airframe: Round-nosed cylindrical body, with four slightly swept flip-out tail-fins.
Guidance and Control: Aeronutronic command guidance system, with infra-red tracking. Control by sustainer exhaust deflection.
Warhead: Octol shaped-charge.
Length: 3ft 9in (1.14m).
Body diameter: 5.95 in (152 mm).
Launch weight: 60lb (27kg).

Development and Service

Philco-Ford claim that Shillelagh was built in greater numbers than any previous US guided missile. It is unusual in being fired from a gun/launcher that can also fire conventional ammunition. This makes it an ideal weapon for tanks and other armoured fighting vehicles, and it has been carried by General Sheridan AFVs of the US Army since 1967, including extensive overseas deployment. Other vehicles equipped with Shillelagh include the American M-60A1E1/E2 battle tank and the new XM-803 (MBT-70) main battle tank.

After firing Shillelagh, the gunner has only to keep the target centred on the cross-hairs of his telescopic sight. An infra-red missile tracking/command system associated with the sight detects and corrects any deviations of the missile's flight path from the line-of-sight to the target. Shillelagh provides greatly increased fire-power for use against all types of armoured fighting vehicles, troops and fortifications. It has been tested in an air-to-surface role from a Bell UH-1B helicopter, in conjunction with a stabilised sight developed by Aeronutronic.

Shrike on A-4 Skyhawk

Shrike (AGM-45A)

USA

Supersonic air-to-surface anti-radar missile. In production and service.

Prime contractor: Naval Weapons Center, China Lake, California.
Powered by: Rocketdyne Mk 39 Mod 7 or Aerojet Mk 53 solid-propellant rocket motor.
Airframe: Slim cylindrical body with ogival nose. Pivoted cruciform wings mid-way from nose to tail, indexed in line with cruciform tail-fins.
Guidance and Control: Passive homing head by Texas Instruments. Control via pivoting wings.
Warhead: High-explosive.
Length: 10ft 0in (3.05m).
Body diameter: 8in (20.3cm).
Wing span: 3ft 0in (0.91m).
Launch weight: 400lb (182kg).
Max range: 10 miles (16km).

Development and Service

Shrike was developed initially under the acronym ARM (anti-radar missile), to home on enemy radars. The need for such a weapon was underlined by the war in Vietnam, where it became increasingly important to knock out or jam the North's warning and missile/aircraft guidance radars during the US air offensive. Delivery to US Navy carrier-based attack squadrons began in 1964. USAF squadrons also adopted Shrike as a standard penetration aid, and it became operational in Vietnam in 1966. Production for the USAF and US Navy continues, and many improvements have been introduced through the years to increase the weapon's effectiveness. One version supplied to the Israeli Air Force is said to have a seeker head designed specifically to home on the radars supplied to Egypt with Soviet 'Guideline' and 'Goa' missile systems.

Sidewinder 1A (AIM-9) — USA

Air-to-air missile. In service.

Data apply to AIM-9B
Prime contractor: US Naval Weapons Center, China Lake, California.
Powered by: Naval Propellant Plant solid-propellant rocket motor.
Airframe: Slim cylindrical aluminium body, with hemispherical glass nose. Delta-shape cruciform control surfaces at nose, indexed in line with large cruciform tail-fins.
Guidance and Control: Infra-red homing guidance. Control by cruciform foreplanes.
Warhead: High-explosive, weighing 25lb (11.4kg).
Length: 9ft 3½in (2.83m).
Body diameter: 5in (13cm).
Span of tail-fins: 1ft 10in (0.56m).
Launch weight: 159lb (72kg).
Max speed: Mach 2.5.
Max range: 2 miles (3.35km).

Development and Service

Sidewinder is one of the most-produced of all airborne missiles. Firing trials began successfully on September 11, 1953, and very large numbers of the infra-red homing AIM-9B were manufactured subsequently by Philco and General Electric for the US services, the Royal Navy, Royal Canadian Navy, Royal Netherlands Navy, the Air Forces of nine NATO nations, Australia, Japan, the Philippines, Spain, Sweden, Taiwan and other countries. In addition, some 9,000 were produced under licence in Europe, by Bodenseewerk of Germany in association with companies in Denmark, Greece, the Netherlands, Norway, Portugal and Turkey.

The popularity of Sidewinder results from the fact that it is claimed to have fewer than two dozen moving parts and no more electronic components than the average home radio. To overcome limitations experienced during air fighting in Vietnam, two new short-range versions were reported to be under development in 1971. The AIM-9J is described as a modified AIM-9B which Philco-Ford is developing to equip the USAF's McDonnell Douglas F-15. The AIM-9H is being developed to arm the US Navy's Grumman F-14 Tomcat and other types.

Sidewinder 1C (AIM-9) USA

Air-to-air missile. In production and service.

Data for AIM-9D.
Prime contractor: US Naval Weapons Center China Lake, California.
Powered by: Rocketdyne Mk 36 Mod 5 solid-propellant rocket motor.
Airframe: Slim cylindrical body, with tapering and rounded nose. Nose control surfaces are of longer chord than on the 1A version, and the cruciform tail surfaces have increased leading-edge sweep.
Guidance and Control: Raytheon infra-red homing (Motorola semi-active radar guidance on AIM-9C). Control via foreplanes.
Warhead: High-explosive.
Length: 9ft 6½in (2.91m).
Body diameter: 5in (13cm).
Span of tail-fins: 2ft 1in (0.64m).
Launch weight: 185lb (84kg).
Max range: over 2 miles (3.5km).

Development and Service
Sidewinder 1C was developed in two versions to offer higher speed and greater range than the 1A series. The AIM-9C radar-homing version is in production by Motorola. The infra-red AIM-9D is manufactured by Raytheon for the US Navy and UK, but is not yet used by the USAF. Sidewinder 1C is operational also in the US Army's Chaparral surface-to-air missile system (see page 25).

Skean (SS-5)   USSR

Intermediate-range ballistic missile. In service.

Powered by: Liquid-propellant rocket motor.
Airframe: Cylindrical body of constant basic diameter, with flared skirt at tail and blunted conical nose-cone. No fins or wings.
Warhead: Nuclear or thermonuclear.
Length: 75ft 0in (23.0m).
Max body diameter: 8ft 0in (2.40m).
Max range: 2,000 miles (3,200km).

Development and Service

Although infinitely more formidable than V-2, 'Skean' can probably be regarded as the ultimate post-war development of the basic concept of that wartime German missile. It appears to be a scale-up of the 'Shyster-Sandal' family, with storable liquid-propellant rocket engine, and is estimated to have a thermonuclear warhead of one megaton yield.

'Skean' was first included in a parade through Moscow in November 1964. Subsequently, Soviet films showed missiles of this type housed in underground silo-launchers, but deployment seems to have been limited, with no more than 100 'Skeans' operational in 1971. Like 'Sandal', these are probably divided between Russia's missile forces aimed at Western Europe and China and Japan.

SLAM

UK

Ship-launched missile system. Under development.

Prime contractor: Vickers Ltd, Shipbuilding Group.

Development and Service

The acronym SLAM signifies either submarine-launched air missile or surface-launched air missile system. In its initial form, the weapon system is being developed for compatibility with the *Oberon* class of submarines; but SLAM could be retro-fitted or built into almost any type of underwater or light surface craft for short-range defence against aircraft, helicopters and enemy surface craft. It consists of a cluster of six Blowpipe missile launchers (see page 20) surrounding an electronic package on a pylon mounting. Equipment in the electronic package includes a TV camera, gyro-stabiliser and some missile control equipment. After launch, each missile is guided by joystick by an operator in the ship's control room, who keeps both missile and target centred on a TV monitor. Preliminary trials of SLAM had been completed by January 1970.

Snapper   USSR

Light anti-tank missile. In service.

Powered by: Solid-propellant rocket motor.
Airframe: Short cylindrical body with conical nose-cone. Rear-mounted cruciform wings of delta shape, each with one or two vibrating spoilers mounted in its trailing-edge.
Guidance and Control: Wire-guided and spin-stabilised. Control by means of spoilers.
Warhead: Hollow-charge, weighing 11.5lb (5.25kg), with contact fuse. Capable of penetrating 13.7in (35cm) of armour.
Length: 3ft 8½in (1.13m).
Body diameter: 5.5in (0.14m).
Wing span: 2ft 5½in (0.75m).
Launch weight: 49lb (22.25kg).
Cruising speed: 120mph (323km/h).
Max range: 7,650ft (2,330m).

Development and Service
Like 'Guideline' and 'Atoll', this is one of the better-known Soviet missiles because examples were captured by the Israeli Army during the 1967 June War. 'Snapper' is a first-generation anti-tank weapon, similar in configuration to Western missiles such as the Nord SS.10 and MBB Cobra. In its initial operational form, as captured by Israel, 'Snapper' was carried on a quadruple launcher by a GAZ-69 jeep-type vehicle. Control was by joystick, and the operator kept the missile aligned on target during flight with the aid of tracking flares mounted on two of the wings. Standard transport/launch vehicle in the Soviet Army is now the armoured, amphibious BRDM, which carries a retractable three-round launcher.

Sparrow launching from an F-4B Phantom II

Sparrow (AIM-7)　　　　　　　　　　　　USA

Long-range air-to-air missile. In production and service.

Prime contractor: Raytheon Company.
Powered by: Rocketdyne Mk 38 Mod 2 solid-propellant rocket motor.
Airframe: Slim cylindrical body with pointed ogival nose. Pivoted cruciform delta wings mid-way along body, indexed in line with cruciform delta tail-fins.
Guidance and Control: Raytheon continuous-wave semi-active radar homing system. Control by movable wings.
Warhead: High-explosive, weighing 60lb (27kg).
Length: 12ft 0in (3.66m).
Body diameter: 8in (20cm).
Wing span: 3ft 4in (1.02m).
Launch weight: 450lb (204kg).
Max speed: above Mach 3.5.
Max range: over 8 miles (13km).

Development and Service
Sparrow is one of the most important missiles currently in service with the NATO air forces, their allies and friends. It is a large, all-weather weapon, of which up to six can be carried by the McDonnell Douglas F-4 Phantom IIs of the USAF, US Navy, US Marine Corps, RAF, Royal Navy and other air forces. The Lockheed F-104S Starfighters of the Italian Air Force each carry two Sparrows, and similar weapons will arm the McDonnell Douglas F-15 and Grumman F-14 Tomcat, America's primary next-generation fighters. Mitsubishi is producing more than 1,000 in Japan, for aircraft of the Japanese Air Self-Defence Force.

The basic current version of Sparrow is the AIM-7E, with the advanced AIM-7F soon to supersede it in production. Sparrow is also used in a ship-based role in the Sea Sparrow weapon system (see page 126).

Spartan (XLIM-49A) USA

Long-range anti-ballistic missile. In production.

Prime contractor: Western Electric Company Inc.
Powered by: Thiokol TX-135, TX-238 and TX-239 solid-propellant rocket motors respectively in the three stages.
Airframe: Tandem three-stage missile. Basically cylindrical body of constant diameter, tapering on final stage towards pointed ogival nose-cone. Cruciform foreplanes on nose. Cruciform delta tail-fins on second stage, indexed in line with large tail-fins on first stage; both indexed at 45 degrees to foreplanes.
Guidance and Control: Bell Telephone Laboratories radar command guidance.
Warhead: Nuclear.
Length: 55ft 0in (16.75m).
Weight and Performance: Secret.

Development and Service

Spartan is the long-range partner to Sprint (page 140) in America's Safeguard ABM defence system (see page 104). Its development dates back to October 1965, when Douglas Aircraft Company received a contract for what was then known as an 'improved Zeus'. It was able to utilise experience gained with the original Nike Zeus experimental ABMs, with the result that firing trials of Spartan from Kwajalein Atoll in the Pacific were able to start in March 1968. An outstanding success was achieved on August 28, 1970, when a Spartan with an inert warhead intercepted at a height of about 100 miles (160km) a Minuteman ICBM re-entry vehicle launched from Vandenberg AFB, California, some 4,200 miles (6,760km) away.

In the operational Safeguard system, the whole sequence from detection of an incoming warhead to its interception by Spartan is intended to be automatic, although manual override will be possible at any stage. The missile's nuclear warhead will be triggered by command at the optimum moment to ensure target destruction above the atmosphere.

Full-size mock-up of Spartan

Sprint

Close-range anti-ballistic missile. In production.

Prime contractor: Martin Marietta Corporation.
Powered by: Hercules Inc solid-propellant rocket motors.
Airframe: Two-stage missile. Conical body, with small cruciform vanes mid-way from nose to tail.
Guidance and Control: Western Electric/Honeywell command guidance and control systems.
Warhead: Nuclear.
Length: 27ft 0in (8.23m).
Diameter at base: 4ft 6in (1.37m).
Launch weight: 7,500lb (3,400kg).

Development and Service

Sprint is the close-range weapon intended to provide a 'last-ditch' defence against enemy ICBMs which elude Spartans (page 139) in America's Safeguard anti-ballistic missile (ABM) system. It is 'popped' from its underground launcher by a separate charge which is placed underneath it when the Sprint is loaded. Its own motor does not ignite until it is clear of the surface; but Sprint then has the highest acceleration of any US missile, and the degree of manoeuvrability essential to intercept an incoming thermonuclear warhead at low altitude. Fine adjustments to the missile's trajectory are made by command signals from ground radar.

Development of Sprint began in 1963. The first firing, in November 1965, was successful; and a major achievement was recorded on December 23, 1970, when a Sprint launched from Kwajalein Atoll in the Pacific intercepted the re-entry vehicle of a Minuteman ICBM fired from Vandenberg AFB, California, about 4,200 miles (6,760km) away. Sufficient Sprints to equip three Safeguard sites were ordered under a production contract placed in December 1970.

SRAAM 75

UK

Short-range air-to-air missile. Under development.

Prime contractor: Hawker Siddeley Dynamics Ltd.
Airframe: Slim cylindrical body with six small rectangular tail surfaces forward of nozzle. Hemispherical glass nose.
Guidance and Control: Infra-red guidance system. Reported to use a thrust-vectoring control system.
Warhead: High-explosive.
Length: 9ft 2in (2.80m).
Body diameter: 6.5in (17cm).

Development and Service

Known originally as Taildog, this weapon has been projected for the same 'dogfight' use as the US Navy's Agile (page 7). Its designation means 'short-range air-to-air missile 75' and it is intended to complement rather than replace weapons such as Red Top. Work was initiated in 1970 under a pre-development contract, with the object of producing a highly-reliable, low-cost weapon compatible with any type of interceptor, strike or reconnaissance aircraft and without requiring aircraft modification of any sort.

Full-size models of SRAAM

SRAM (AGM-69A; WS-140A) USA

Air-to-surface short-range missile. In production.

Prime contractor: The Boeing Company.
Powered by: Lockheed Propulsion Company LPC-415 restartable two-pulse solid-propellant rocket motor.
Airframe: Body of circular section, tapering at tail and fitted with long ogival pointed nose-cone. Three equi-spaced tail control surfaces around nozzle, which is faired off by a destructible tail-cone when the missile is carried externally.
Guidance and Control: Inertial guidance system by Kearfott Systems Division of General Precision Inc. Control by tail surfaces.
Warhead: Nuclear.
Length: 14ft 0in (4.27m).
Body diameter: 1ft 5½in (44.5cm).
Weight (approx.): 1,985lb (900kg).
Cruising speed: Mach 2.5.
Max range: 100 miles (160km).

SRAM eight-round rotary launcher

Development and Service

Boeing started work on a short-range attack missile (SRAM), to extend the effective life of the B-52G and H versions of their Stratofortress strategic bomber, in December 1963. By July 1965, the USAF had formulated a need for a weapon of this type; after studying tenders from five competing companies, they selected Boeing as prime contractor for the AGM-69A (Weapon System 140A). Present plans are for each B-52G/H to carry 20 SRAMs, of which twelve will be in the form of three-round underwing clusters and eight will go in the aft bomb-bay, together with four Mk 28 thermonuclear weapons. The swing-wing FB-111As of two Strategic Air Command wings will each be able to carry six SRAMs, four on swivelling underwing pylons and two internally. Development launches from both types of aircraft have been under way since 1969. In one test by a USAF crew from a B-52 in 1970, two missiles launched in rapid succession hit two separate targets. Production was authorised by the USAF in January 1971, and the SRAM will also arm the North American Rockwell B-1 supersonic bomber if this reaches service status.

Key to the accuracy of SRAM is the information fed into its guidance system before launch. The Guidance and Controls Division of Litton Industries is responsible for the B-52 inertial measurement unit; Autonetics Division of North American Rockwell for the B-52/FB-111 aeroplane computer; IBM for modifying the B-52 bomb-nav system. The missile can fly 'dog-leg' courses to elude the defences, is unaffected by enemy ECM and appears no larger than a machine-gun bullet on radar screens. Its range is said to vary from 100 miles (160km) at high altitude to 35 miles (55km) at low level.

SRAM check-out before delivery

SRAM launch from a B-52

SS.10 (Nord 5203)

France

Anti-tank missile. In service.

Prime contractor: Nord-Aviation.
Powered by: Two-stage solid-propellant rocket motor.
Airframe: Cylindrical body with rounded ogival nose. Rear-mounted cruciform wings, each with a small control surface at the trailing-edge root.
Guidance and Control: Wire-guided. Control by wing-root spoilers.
Warhead: High-explosive, weighing 11lb (5kg).
Length: 2ft 9¾in (0.86m).
Body diameter: 6.5in (16.5cm).
Wing span: 2ft 5½in (0.75m).
Launch weight: 32.6lb (14.8kg).
Max speed: 177mph (285km/h).
Range limits: 990-4,920ft (300-1,500m).

Development and Service

Contemporary with Entac (page 31), and similar in configuration and size, the SS.10 was one of the first guided missiles to be used in action. The Israeli Army employed it successfully against Egyptian armoured fighting vehicles during the Suez campaign of 1956. Many other countries adopted it as a standard infantry weapon, including France, Sweden and West Germany, and production was at the rate of 450-500 per month for a long period. At least nine countries still included SS.10s in their combat inventory at the start of the 'seventies, although the weapon has long been superseded by newer types. It was designed to be fired normally from the ground, using its boxlike container as launcher. It was also fired effectively from jeeps, light aircraft and helicopters.

SS.11s on Jeep launcher

SS.11 — France

Surface-to-surface battlefield missile. In production and service.

Prime contractor: Nord-Aviation/Aérospatiale, Division Engins Tactiques.
Powered by: Two-stage solid-propellant rocket motor.
Airframe: Cylindrical body, with sweptback cruciform wings and rounded ogival nose.
Guidance and Control: Wire guidance by line-of-sight command via operator's joystick controller. Spin-stabilised. Control by deflection of the sustainer efflux.
Warhead: Alternative armour-piercing type able to penetrate 24in (600mm) of steel, high-explosive type which will penetrate 0.4in (10mm) of steel plate at full range, or high-fragmentation anti-personnel type with contact fuse.
Length: 3ft 11in (1.20m).
Body diameter: 6.5in (16cm).
Wing span: 1ft 7½in (0.50m).
Launch weight: 66lb (29.9kg).
Cruising speed: 360mph (580km/h).
Range limits: 1,650-9,840ft (500-3,000m).

Development and Service
Nord-Aviation developed this very effective tactical missile as a follow-on to its success with the SS.10 and Entac. Production has been continued by Aérospatiale, at the rate of 600 a month (including AS.11s), and a total of 148,000 missiles had been delivered by 1971. They are used by all three French Services and by the armed forces of 18 other countries including the USA and UK.

Being larger than the SS.10, the SS.11 is fired normally from ground vehicles and light naval vessels, including hovercraft; but it can be dismounted for operation from a ground launcher by mountain combat, airborne and coastal defence forces. In action, the operator follows both the target and the missile by means of a magnifying optical sight, and guides the missile to impact by means of a control joystick. Tracking is facilitated by flares on the rear of the missile. The standard production version since 1962 has been the SS.11 B.1 (and AS.11 B.1) with transistorised firing equipment. The Harpon (page 50) is similar except for its use of TCA guidance.

SS-11 — USSR

Intercontinental ballistic missile. In service.

Powered by: Storable liquid-propellant rocket engines.
Warhead: Thermonuclear, of one megaton yield, with radar decoys. Alternatively, a multiple re-entry vehicle (MRV), with three separate charges and decoys, may be operational.
Max range: 6,525 miles (10,500km).

Development and Service
Little is known about the SS-11, although official US statements suggested that 800 of the 1,200 Soviet ICBMs emplaced in 1970 were of this type. It is said to have been operational in camouflaged silo launchers since 1966, and to be due for replacement by the solid-propellant SS-13. US monitoring services reported that two SS-11s, with warheads equivalent to a one-megaton charge, were launched 5,715 miles (9,200km) from Tyuratam test site in Kazakhstan into the Pacific on July 27 and 28, 1970. Two more SS-11s were identified at the end of 6,525 mile (10,500km) flights with MRV warheads on August 21, 1970.

SS.12

France

Surface-to-surface tactical missile. In service.

Prime contractor: Nord-Aviation/Aérospatiale, Division Engins Tactiques.
Powered by: Dual-thrust solid-propellant rocket motor.
Airframe: Cylindrical body with mid-positioned cruciform wings and bulged ogival nose. Two sustainer nozzles aft of wings, on opposite sides of body.
Guidance and Control: Wire-guided and spin-stabilised. Alternatively, the TCA optical aiming/infra-red tracking system can be used. Control by sustainer jet deflection.
Warhead: Alternative armour-piercing type capable of penetrating more than 1.5in (40mm) of steel, shaped charge for anti-tank use, or pre-fragmented anti-personnel type, each weighing 66lb (30kg).
Length: 6ft 1.9in (1.87m).
Body diameter: 7in (18cm).
Wing span: 2ft 1½in (0.65m).
Launch weight: 167lb (75kg).
Cruising speed: 425mph (685km/h).
Max range: 19,650ft (6,000m).

Development and Service
By scaling up their very successful SS.11, Nord-Aviation (now Aérospatiale) were able to produce the still-compact SS.12 with a warhead about four times as heavy as that of the smaller missile. This makes it suitable for use against fortifications as well as all types of vehicles, including tanks and ships. The standard version uses optical line-of-sight command guidance, through trailing wires; but the SS.12 can be adapted to utilise the TCA semi-automatic type of guidance system, as described for the AS.30 air-to-surface weapon system (see page 13). Another derivative is the 'marine' SS.12M, which was first demonstrated to naval officers of eight countries from the coastal patrol boat *La Combattante,* off Toulon, on June 15, 1966. On that occasion, two SS.12Ms were fired from a lightweight launcher, with self-contained optical sighting, guidance and control equipment, against a moving target nearly 3½ miles (5.5km) away. Both impacted the target within 3ft (1m) of dead centre and the same distance above the waterline. First operational fast patrol boats to be fitted with this weapon system were three Vosper Thornycroft ships of the Libyan Navy. Each carries two four-round launchers able to mount SS.12Ms for combat use or SS.11 for training.

SSBS

France

Surface-to-surface ballistic missile. In service.

Prime contractor: Aérospatiale, Division des Systèmes Balistiques et Spatiaux.
Powered by: First stage: PNSM P.16 (Norma Type 902) solid-propellant motor. Second stage: PNSM P.10 (Norma Type 903) solid-propellant motor.
Airframe: Two-stage missile. Cylindrical body of constant diameter, with basically cylindrical re-entry vehicle of smaller diameter. First-stage casing of maraging metal; second-stage casing of Vascojet 1000. No fins or wings.
Guidance and Control: Inertial guidance system. Each stage has four gimballed nozzles for control.
Warhead: Nuclear, approx 150-kiloton yield.
Length: 48ft 6½in (14.80m).
Max body diameter: 4ft 11in (1.50m).
Launch weight: 70,327lb (31,900kg).
Max range: 1,865 miles (3,000km).

Development and Service

Two operational squadrons of the 1er Groupement de Missiles Stratégiques of the French Air Force deploy a total of 18 SSBS (sol-sol balistique stratégique) missiles. These are housed in underground silos on the Plateau d'Albion, east of Avignon in Haute Provence, with fire control centres at Rustrel (Vaucluse) and Reilhannette (Drôme). The missiles are held at a state of readiness and are designed to be launched by the rapid remotely-controlled opening of the silo doors, without requiring any human action at the widely-dispersed launch areas.

It was intended originally to form three squadrons of SSBS missiles as the second-generation 'force de frappe', supplementing Mirage IVA strategic bombers of the French Air Force and with the French Navy's MSBS submarines to follow. Budget cutbacks compelled the deletion of one squadron. Deployment of the others started after the final S-02

SSBS inside silo launcher

prototype missile made a fully successful 1,550-mile (2,500-km) flight towards the Azores, from a silo at Landes test centre, on June 24, 1969. The present warheads will be replaced with thermonuclear warheads in due course.

SS-N-?

USSR

Surface-to-surface missile. In service.

Development and Service

A new short-range surface-to-surface missile is carried by the five nuclear-powered submarines of the Soviet Navy's high-speed 'C' class, each of which has eight launch-tubes. Estimated range of this missile, which does not yet have a western designation or NATO code-name, is 30 miles (48km).

Standard ARM (AGM-78)

USA

Air-to-surface and surface-to-surface anti-radiation missile. In production and service.

Prime contractor: General Dynamics Corporation, Electro Dynamics Division.
Powered by: Aerojet-General dual-thrust solid-propellant rocket motor.
Airframe: Cylindrical body with pointed ogival nose. Cruciform long-chord narrow-span wings, with forward portions of much reduced span, indexed in line with cruciform tail control surfaces.
Guidance and Control: Passive homing guidance system, using seeker head which homes on enemy radar emissions. Control by tail surfaces.
Warhead: High-explosive.
Length: 15ft 0in (4.57m).
Body diameter: 1ft 0in (30.5cm).
Launch weight: 1,800 lb (816kg).
Max speed: Mach 2.
Max range: 15.5 miles (25km).

Development and Service

Design studies for this anti-radiation version of the RIM-66A Standard Missile (see page) were started in 1966. Flight trials followed in 1967 and the weapon entered full production in 1968. Its purpose is to home on and destroy enemy defensive and missile guidance radars. This must be possible in an ECM environment, and with the knowledge that an enemy operator will switch off a radar the moment he suspects that its signals are being picked up and used to provide target data for attack. As a result, a key item of support equipment is the Target Identification and Acquisition System (TIAS) carried by aircraft used as launch vehicle for Standard ARM. This is able to compute a trajectory for the missile on the basis of data already monitored in the event of the target radar being switched off.

Standard ARM is carried by the US Navy's

Standard ARM launcher on US Navy gunboat

Standard ARM under wing of A-6A Intruder

A-6 and USAF's F-105, and is intended for deployment on other types such as the E-2C Hawkeye and EA-6B ECM aircraft. The original version used a standard Texas Instruments seeker head, as fitted to the Shrike missile. This was followed by an improved seeker developed by Maxson Electronics on the AGM-78B. Further-improved versions are being developed, with a seeker by GD/Pomona.

The US Navy carried out firing trials of Standard ARM missiles from a special box-launcher mounted on a patrol gunboat during 1971, to evaluate the usefulness of the weapon system in a surface-to-surface role. The results are believed to have been successful.

Standard Missile (RIM-66A/RIM-67A)   USA

Supersonic ship-launched surface-to-air missiles. In production and service.

RIM-66A on shipboard launcher

Prime contractor: General Dynamics Corporation, Electro Dynamic Division.
Powered by: RIM-66A has dual-thrust solid-propellant rocket motor. RIM-67A has tandem two-stage solid-propellant rocket motors. All motors are produced by the Navy Ordnance System Command.
Airframe: RIM-66A has cylindrical body with pointed ogival nose, and cruciform long-chord narrow-span wings, with forward portions of much reduced span, indexed in line with cruciform tail control surfaces. Second stage of RIM-67A is externally similar to RIM-66A; first stage comprises a cylindrical booster with cruciform tail-fins indexed in line with second-stage surfaces.
Guidance and Control: General Dynamics/Pomona semi-active radar homing guidance. Control by tail surfaces.
Warhead: High-explosive, with impact and proximity fuses.
Length: RIM-66A 15ft 0in (4.57m).
RIM-67A 27ft 0in (8.23m).
Body diameter: 1ft 0in (30.5cm).
Launch weight: RIM-66A 1,300lb (590kg).
RIM-67A 3,000lb (1,360kg).
Max range: RIM-66A 15 miles (24km).
RIM-67A 35 miles (56km).

Development and Service

These two ship-launched missiles were developed by General Dynamics to replace their Tartar and Advanced Terrier, which the new weapons resemble externally. Simplicity and reliability result from the switch to solid-state electronics and an all-electric control system, with no pneumatics or hydraulics. The two configurations, known sometimes as Medium-range Standard Missile (MR) and Extended-range Standard Missile (ER), are intended for use on cruisers and destroyers (MR) and frigates (ER). They are effective against both aircraft and anti-ship cruise missiles, down to very low levels, and have a surface-to-surface capability. About 70 ships of the US Navy had been refitted with Standard Missile by mid-1971. Deployment is scheduled for completion by the mid-70s. Standard Missile forms the basis of the Standard ARM anti-radiation missile system. It will also be embodied in the initial Aegis defence system (see page 6).

Styx (SS-N-2)

USSR

Surface-to-surface ship-launched cruise missile. In service.

Powered by: Nozzle for sustainer rocket motor in tail. Jettisonable solid-propellant booster, with downward-canted nozzle, under rear fuselage.
Airframe: Aeroplane configuration. Circular-section body, with rounded nose-cone and tapered towards tail. Mid-set cropped-delta wings, with ailerons of almost full span. Three identical tail surfaces, each made up of a fixed fin and trailing-edge control surface.
Guidance and Control: Reported to use radar homing system. Control via ailerons and tail control surfaces.
Warhead: High-explosive.
Length: 21ft 4in (6.5m).
Max body diameter: 2ft 3in (0.70m).
Wing span: 8ft 10in (2.70m).
Max range: 29 miles (46km).

Development and Service

This ship-to-ship weapon has the distinction of being one of the few modern surface-to-surface missiles of proven operational effectiveness. Its opportunity came on October 21, 1967, when the Israeli destroyer *Eilat* was detected sailing off Port Said, at a range of around 15 miles (24km). A series of 'Styx' missiles were launched from Soviet-built *Osa* class patrol boats of the Egyptian Navy, from the sheltered waters of the harbour. Some of these scored direct hits on the *Eilat,* which was sunk.

Each *Osa* class ship carries four 'Styx' missiles ready for launching in containers to each side of its rear deck. *Komar* class fast patrol boats are fitted with two similar containers. The Soviet Navy has a total of about 125 ships of these classes. Nearly as many more have been delivered to the Navies of Algeria, Communist China, Cuba, Egypt, East Germany, Indonesia, Poland, Romania, Syria and Yugoslavia. In addition, the six Soviet corvettes of the *Nanuchka* class each carry six 'Styx' missiles in two triple launchers.

Subroc (UUM-44A)   USA

Underwater-to-underwater missile. In production and service.

Prime contractor: Goodyear Aerospace Corporation.
Powered by: Thiokol solid-propellant tandem booster, with four nozzles embodying jet-deflection.
Airframe: First stage comprises the Thiokol rocket motor in its cylindrical casing. The second stage is also cylindrical, with a short ogival nose-cone and four tail-fins, and is of smaller diameter than the first stage.
Guidance and Control: Lightweight inertial guidance system by General Precision Inc. Control by thrust vectoring and aerodynamic tail-fins.
Warhead: Nuclear depth bomb.
Length: 21ft 0in (6.40m).
Max body diameter: 1ft 9in (0.53m).
Launch weight: 4,000lb (1,815kg).
Max range: 25-30 miles (40-48km).

Development and Service

Subroc is one of the most unusual missiles in service. Fired from the standard torpedo tube of a submerged submarine, it is designed to propel itself to the surface, then follow a ballistic trajectory to the predicted position of an enemy submarine, which it attacks with a nuclear depth bomb warhead. To make this possible, the large booster exhausts through thrust-vectoring nozzles which enable the missile to change course underwater as well as in flight. The motor does not ignite until the missile has travelled a safe distance from its launch submarine, which need not be pointing towards the target. The Subroc is propelled upward and out of the water. Its inertial guidance system then directs it towards the enemy craft, whose position, course and speed have been computed and fed into the missile guidance before launch. At a pre-determined range, the booster is separated by thrust reversal and a mechanical disconnect system. The depth bomb continues in the ballistic trajectory, steered by its tail-fins in accordance with instructions from the inertial guidance system. This determines the position and angle of the missile as it re-enters the water. A special mitigating device cushions the shock of water entry at supersonic speed; the depth bomb then sinks and explodes.

Subroc became operational in 1965, after a highly successful series of full-range tests in the Pacific from the USS *Plunger*. Production continues, to arm high-speed nuclear attack submarines of the US Navy in world-wide service.

Super 530
France

Air-to-air missile. Under development.

Prime Contractor: SA Engins Matra.

Development and Service
Little information is yet available on this new weapon which will arm Dassault Mirage F1 interceptors of the French Air Force. It is said to embody components of the R.530 (see page —) but with improvements to the aerodynamic design, power plant and electronics which will more than double the effective range. All that has been stated officially is that the Super 530 will be an all-weather missile with the capability of destroying targets flying at a height very different from that of the launch aircraft.

Super Falcon (AIM-4F/G)
USA

Air-to-air missile. In service.

Prime contractor: Hughes Aircraft Company.
Powered by: Thiokol M46 two-stage solid-propellant motor, with first-stage rating of 6,000lb (2,720kg) st.
Airframe: Cylindrical body, with cruciform long-chord delta wings, each extended at leading-edge root to form a narrow strake along the fore-part of the body. Cruciform tail control surfaces aft of wings. AIM-4F has ogival nose with central probe; AIM-4G has tapered nose with hemispherical glass tip.
Guidance and Control: AIM-4F utilises Hughes semi-active radar homing guidance; AIM-4G uses infra-red homing system. Control by tail surfaces.
Warhead: High-explosive, weighing 40lb (18kg).
Length: AIM-4F 7ft 2in (2.18m).
AIM-4G 6ft 9in (2.06m).
Body diameter: 6.6in (17cm).
Wing span: 2ft 0in (0.61cm).
Launch weight: AIM-4F 150lb (68kg).
AIM-4G 145lb (65.7kg).
Max speed: Mach 2.5.
Max range: 7 miles (11km).

Development and Service
Super Falcon was developed as an interim stage between the AIM-4A/C Falcon and the AIM-26 Falcon. First version, introduced in 1958, was the radar-homing AIM-4E (originally GAR-3), which was succeeded in 1960 by the improved AIM-4F (GAR-3A) after only 300 Es had been built. Simultaneously, Hughes introduced the AIM-4G (GAR-4A), with infra-red homing head, and a mixed armament of four AIM-4F/Gs has since been standard in the internal weapons bay of USAF Convair F-106 Delta Dart interceptors. As well as having a higher performance than early Falcons, they are less susceptible to enemy countermeasures.

AIM-4F (*left*) and AIM-4G Super Falcon

Swatter

USSR

Anti-tank missile. In service.

Powered by: Solid-propellant motor.
Airframe: Cylindrical body with rounded nose. Cruciform wings at rear, each with trailing-edge elevon. Tracking flare inboard of elevon on two wings. Two small movable foreplanes. Two motor nozzles on opposite sides of body between wings.
Guidance and Control: Wire guidance, possibly with terminal homing. Control via elevons, possibly supplemented by foreplanes during terminal homing.
Warhead: High-explosive.
Length: 3ft 8in (1.12m).
Wing span: 2ft 2in (0.65m).

Development and Service
This second-generation missile was first observed on a quadruple, retractable launcher carried by a BRDM vehicle, some years after 'Snapper' entered service. 'Swatter' is a more advanced weapon, possibly in the class of the French SS.11.

Swingfire

Long-range anti-tank missile. In production and service.

Prime contractor: British Aircraft Corporation, Guided Weapons Division.
Powered by: Solid-propellant rocket motor.
Airframe: Cylindrical body, with ogival nose and spring-loaded cruciform wings which fold against the rear of the body while the missile is in its launcher.
Guidance and Control: Wire-guidance by command signals transmitted by operator via joystick controller. Control by gimballed motor nozzle, known as a jetavator.
Warhead: Hollow-charge high-explosive warhead, capable of destroying any known tank.
Length: 3ft 6in (1.07m).
Body diameter: 6.7in (17cm).
Wing span: 1ft 2.7in (0.37m).
Max range: 13,125ft (4,000m).

Development and Service

Before the weapon activities of Fairey Engineering Ltd were transferred to British Aircraft Corporation (AT) Ltd, Fairey was working on a long-range command-guided anti-tank weapon known as Orange William. Features of this missile were incorporated in Swingfire, of which development began in the early 'sixties to replace the Malkara in British Army service. Its name reflects the fact that it has a firing arc of 90 degrees from a fixed launcher, enabling it to be launched around corners from a concealed vehicle. When mounted on a traversing turret, targets can be engaged through a full 360-degree field of fire.

The rocket motor is designed to impart relatively low launch acceleration, making Swingfire sufficiently manoeuvrable to engage targets over a direct-fire range of under 500ft (150m). It is launched from its sealed disposable storage container, which can be mounted on a variety of vehicles, such as the FV 438 armoured personnel carrier, which carries two missiles ready to fire and twelve more stowed; the FV 712 Ferret Mk 5 scout car, which carries four ready to fire and two spares; and the British Army's new Alvis Striker tracked reconnaissance combat vehicle. Regiments of the Royal Armoured Corps have been operational with Swingfire since 1969. The Belgian Army will also use the missile on Striker vehicles; and an air-to-surface version is being developed for use from helicopters such as the Lynx and Gazelle. A six-round palletised version of Swingfire can be deployed on unmodified vehicles such as the long-wheelbase Land Rover, and is suitable for helicopter transportation, air-dropping and operation from ground emplacements; its total weight is only 1,000lb (453kg).

Talos (RIM-8G-AAW and RIM-8H-ARM) — USA

Ship-launched surface-to-air and surface-to-surface missile. In service.

Prime contractor: The Bendix Corporation, Aerospace Systems Division.
Powered by: 40,000 hp Bendix 28in (710mm) ramjet sustainer and Allegany Ballistics tandem jettisonable solid-propellant booster.
Airframe: Tandem two-stage missile. Cylindrical missile body, tapering towards nose air-intake which contains conical centre-body. Pivoted cruciform wings mid-way back on body and indexed in line with rectangular cruciform tail-fins. Cylindrical booster with four tail-fins indexed in line with fins of missile.
Guidance and Control: Beam-riding, with terminal semi-active radar homing. Control by pivoted wings.
Warhead: Alternative nuclear or high-explosive, with proximity fuse.
Length: With booster 38ft 0in (11.58m). Without booster 21ft 0in (6.40m).
Body diameter: 2ft 4in (71cm).
Wing span: 9ft 6in (2.90m).
Launch weight: 3,400lb (1,542kg).
Max speed: Mach 2.5.
Max range: 70 miles (112km).

Development and Service
The effectiveness of this big ramjet-powered missile was demonstrated in 1968, when Talos missiles launched from the nuclear-powered cruiser *Long Beach* destroyed two MiG fighters flying over North Vietnam, some 70 miles (112km) from the ship. Five other US Navy cruisers continue to deploy Talos, and the designation RIM-8H-ARM given to one current version suggests that it carries an anti-radiation homing head for surface-to-surface use against enemy radars at sea or on land.

Talos, like Terrier (page 5), was one of the first US surface-to-air missiles, tracing its development back to the Bumblebee research programme which the Applied Physics Laboratory of John S Hopkins University initiated in 1944. It was first fired at sea, from the USS *Galveston*, on February 24, 1959.

Tartar (RIM-24)

USA

Ship-based surface-to-air missile. In service.

Prime contractor: General Dynamics Corporation.
Powered by: Aerojet-General dual-thrust solid-propellant rocket motor.
Airframe: Cylindrical body with ogival nose, cruciform long-chord wings of constant narrow span and cruciform tail control surfaces immediately to rear of wings.
Guidance and Control: Raytheon semi-active radar homing. Control by tail surfaces.
Warhead: High-explosive, with impact and proximity fuses.
Length: 15ft 0in (4.57m).
Body diameter: 1ft 1.4in (34cm).
Launch weight: over 1,200lb (545kg).
Max speed: above Mach 2.5.
Max range: over 10 miles (16km).

Development and Service

In general appearance Tartar resembles the second stage of the Advanced Terrier missile (page 5). It was developed by the Bureau of Naval Weapons, and produced by Convair Pomona, to provide a compact primary anti-aircraft defence for US Navy destroyers and destroyer escorts, and secondary defence for cruisers. Full production began in 1960 and was completed in 1968, by which time Tartar equipped 33 US Navy ships, and three Australian, four French, one Japanese and two Italian destroyers. Tartar is being replaced by RIM-66A medium-range Standard Missile, which resembles it in appearance.

Terne Mk 8

Norway

Anti-submarine rocket depth-charge. In service.

Prime contractor: A/S Kongsberg Vaapenfabrikk.
Powered by: Two concentric solid-propellant rocket motors.
Airframe: Cylindrical body with cruciform tail-fins. Ogival nose-cone, tapering to slim cylindrical tip.
Guidance and Control: Sonar on ship determines range, bearing and depth of target before Terne is fired. No in-flight guidance.
Warhead: High-explosive, weighing 110lb (50kg), with combined acoustic proximity, impact and time fuse.
Length: 6ft 4¾in (1.95m).
Body diameter: 8in (20.3cm).
Launch weight: 298lb (135kg).
Performance: Secret.

Development and Service

Operational on ships of the Royal Norwegian Navy, this rocket-propelled depth charge system can be installed on ships of any size down to small escort vessels. Once the ship's sonar has established the position of the enemy submarine, a full salvo of six Ternes can be fired in five seconds. The improved rocket motors of Terne Mk 8 give it about twice the range of the earlier Mk 7 version.

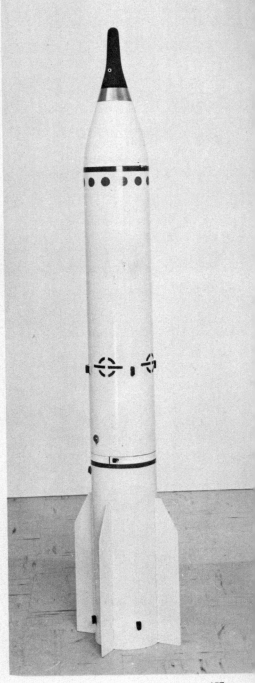

Thunderbird in British Army service

Thunderbird UK

Surface-to-air missile. In service.

Data apply to Thunderbird Mk 2.
Prime contractor: British Aircraft Corporation, Guided Weapons Division.
Powered by: Solid-propellant rocket motor. Four jettisonable solid-propellant boosters.
Airframe: Cylindrical body, with fixed cruciform wings positioned well back and indexed in line with cruciform tail control surfaces. Four boosters wrapped around rear end of body, each with wedge nose and carrying a large stabilising fin. Pointed ogival nose-cone.
Guidance and Control: Semi-active radar homing guidance. Control by tail surfaces.
Warhead: High-explosive, with proximity fuse.
Length: 20ft 10in (6.35m).
Body diameter: 1ft 8¾in (0.53m).
Wing span: 5ft 4in (1.63m).
Weight and Performance: Secret.

Development and Service
The original Mk 1 version of Thunderbird continues to be operated in Saudi Arabia, which acquired 37 ex-British Army missiles and launchers in 1967. The Mk 2 version, which is the British Army's standard heavy anti-aircraft missile, differs in having a longer range, improved low-altitude effectiveness, improved mobility and greater resistance to electronic countermeasures, largely as a result of the switch to a CW (continuous wave) radar. It equips No 36 Heavy Air Defence Regiment of the Royal Artillery, based in Germany. Each Thunderbird battery comprises surveillance and height-finding radars and a battery command post controlling up to six firing troops. Each firing troop has a launch control post, a target illuminating radar and three missile launchers.

BAC offers for export a comprehensive air defence system known as Thunderbird 22, which includes a Plessey Radar/Elliott-Automation Nomad missile and fighter control centre, radars and batteries of Rapier and Thunderbird missiles.

Tigercat UK

Close-range surface-to-air missile system. In production and service.

Prime contractor: Short Brothers and Harland Ltd.

Development and Service
This missile system was evolved specifically for the point defence of airfields and other important potential target areas on land. The Tigercat missile is identical with Seacat (see page 121), which speeded development of the system, and it has been operational with the RAF Regiment since 1970. Other users include the Imperial Iranian Air Force, the Armies of Argentina, Jordan and Qatar, and the Argentinian Marine Corps.

In its mobile form, Tigercat comprises a three-round launcher-trailer and a director trailer, housing the optical sighting and launch control equipment. They can be towed by two vehicles such as Land-Rovers, carrying the launch crew, additional missiles, generator and other items. By integrating the system with surveillance radar, it can be given day and night all-weather capability.

Tigercat launcher-trailer

Titan II (LGM-25C) USA

Intercontinental ballistic missile. In service.

Prime contractor: Martin Marietta Corporation.
Powered by: First stage: Aerojet-General LR87 storable liquid-propellant engine rated at 430,000lb (195,000kg) thrust. Second stage: Aerojet-General LR91 storable liquid-propellant engine rated at 100,000lb (45,360kg) thrust.
Airframe: Two-stage missile. Cylindrical body of high-strength aluminium alloy, with constant diameter. Blunt-nosed conical warhead. No fins or wings.
Guidance and Control: Inertial guidance system by AC Electronics. Control by two gimballed nozzles on first stage and one on second stage.
Warhead: Thermonuclear, in General Electric Mk 6 ablative re-entry vehicle.
Length: 103ft 0in (31.40m).
Max body diameter: 10ft 0in (3.05m).
Launch weight: 330,000lb (149,690kg).
Max speed: 17,000mph (27,360km/h).
Max range: 6,300 miles (10,140km).

Development and Service

The USAF's Strategic Air Command deployed originally two huge liquid-propellant ICBMs, named Atlas and Titan. Titan I became operational in 1962 and was joined by Titan II in the following year. This was by far the most formidable of all America's first-generation long-range missiles. It could be stored in, and launched from, an underground silo, and carried the largest H-bomb warhead that has ever been fitted to a US missile. As a result, the USAF decided to retain six squadrons of Titan IIs, with a total of 54 missiles, although its original intention had been to withdraw all liquid-propellant ICBMs from service as the Minuteman force became operational.

The Titan IIs are based at Davis-Monthan AFB, Arizona, McConnell AFB, Kansas, and Little Rock AFB, Arkansas. The re-entry vehicle carried by each missile is so designed that its speed and trajectory are corrected by four small vernier rockets before it separates from the burned-out second stage. It carries advanced penetration aids to make detection and destruction by an ABM extremely difficult.

TOW (MGM-71A) USA

Heavy anti-tank missile. In production and service.

Prime contractor: Hughes Aircraft Company.
Powered by: Hercules Inc solid-propellant motor, with two separate boost periods.
Airframe: Cylindrical body, tapering to a rounded nose section of smaller diameter. Cruciform wings, mid-way back on body, are folded inside launch-tube and flick forward as missile leaves launcher. Cruciform narrow-chord tail control surfaces, indexed at 45 degrees to wings, flick open rearward at launch.
Guidance and Control: Wire guidance. Control by tail surfaces.
Warhead: High-explosive shaped charge.
Length: 3ft 9.7in (1.16m).
Max body diameter: 5.9in (15cm).
Launch weight: 42lb (19kg).
Max speed: High subsonic.
Range limits: 80-6,600ft (25-2,000m).

TOW three-round launch-packs on UH-1 helicopt

Development and Service

The TOW (Tube-launched Optically-tracked Wire-guided) anti-tank weapon system was conceived for the US Army as a replacement for the 106mm recoilless rifle and Entac and SS.11 missiles. It consists of a glass-fibre launch-tube, a tripod mounting, a traversing and sighting unit, an electronic package and the TOW missile in a shipping/launch container which forms an extension of the launch-tube. Weight of the complete system is about 200lb (91kg), but it can be broken down into four units for carriage by infantry. The missile itself is never handled by the launch-crew, all electrical and mechanical connections being made automatically when its container is inserted in the launch-tube. The solid-propellant motor burns initially only long enough to propel TOW from the launcher. To ensure safety for the operator, the missile then coasts for a time before the second stage fires, well clear of the mouth of the launch-tube. To guide it to the target, the operator has only to keep the target centred in a telescopic sight.

Firing trials were under way successfully in 1965, and production began in November 1968 under a three-year initial contract. Deployment to US Army units in the USA and Europe began in November 1970, and the West German Army selected TOW to replace the MBB Cobra in 1971. A three-round pack and gyro-stabilised sight have been developed for carriage by armed helicopters such as the Bell UH-1 and AH-1, Lockheed AH-56A and Sikorsky Blackhawk. The Bell KingCobra can carry two four-tube packs.

TOW vehicle-mounted launcher

ULMS

USA

Underwater-to-surface ballistic missile. Under development.

Prime contractor: Lockheed Missiles and Space Company.

Development and Service
The ULMS (Undersea Long-range Missile System) has been under study for some years as a third-generation follow-on to Polaris and Poseidon. The intention is to develop a weapon that would have a range of at least 5,750 miles (9,260km). This would make the carrier-submarines much more difficult to locate and track, by extending immensely the area of ocean in which they could patrol and still be within range of their designated targets. It would even be possible to install such a weapon system in short-range submarines, built to cruise along the US continental shelf where they would be difficult to detect by sonar.

A total of $55 million was allocated for initial research and development of the ULMS concept by the US Navy in fiscal years 1970-71. The decision to progress towards an operational system was indicated by the award of a $25 million contract to LMSC, as prime contractor, in November 1971. Initially, the ULMS is expected to replace Poseidon in the launch-tubes of existing submarines.

Vigilant

UK

Lightweight anti-tank missile. In production and service.

Prime contractor: British Aircraft Corporation, Guided Weapons Division.
Powered by: ICI two-stage solid-propellant rocket motor.
Airframe: Cylindrical body, with forward portion of larger diameter than rear section. Conical nose with small-diameter cylindrical projection at centre. Cruciform rectangular wings on rear part of body, each with trailing-edge control-flap.
Guidance: Wire-guidance, with optical line-of-sight command via operator's thumb movements. Control by flaps on wings.
Warhead: Hollow charge, weighing 13.2lb (6.0kg) and able to penetrate more than 22in (558mm) of steel armour.
Length: 3ft 6.2in (1.07m).
Body diameter: 4.5in (11cm).
Wing span: 11in (0.28m).
Launch weight: 31lb (14kg).
Cruising speed: 348mph (560km/h).
Range limits: 750-4,500ft (230-1,375m).

Development and Service
First anti-tank guided weapon developed in the UK, Vigilant is small enough to be carried and fired by one man but sufficiently powerful to knock out the largest known tank. Work on the project was initiated by Vickers-Armstrongs (Aircraft) Ltd (now part of BAC) in 1956 and Vigilant was adopted by the British Army in 1964. It is a standard weapon with British infantry battalions and on Ferret 2/6 scout cars of the Royal Armoured Corps. Other users include the defence forces of Abu Dhabi, Finland, Kuwait, Libya and Saudi Arabia.

Use of an advanced velocity control system, with twin-gyro autopilot, simplifies the operator's task, and the average trainee can hit the target with his first missile after simulator training. The flap control surfaces, instead of the more-usual spoilers, convey high manoeuvrability, contributing to the short-range capability of Vigilant. Guidance is by optical line-of-sight command, via a thumbstick control, tracking being aided by a flare on the missile. A single operator can control up to six missiles.

Walleye   USA

Air-to-surface 'glide bomb'. In production and service.

Prime contractor: Martin Marietta Corporation.
Airframe: Cylindrical body, with tapering tail-cone and glass-tipped ogival nose. Rear-mounted cropped-delta cruciform wings, each with trailing-edge control surface.
Guidance and Control: Self-homing TV guidance system. Control by aerodynamic surfaces on wings.
Warhead: High-explosive.
Length: 11ft 3in (3.43m).
Body diameter: 1ft 3in (38cm).
Wing span: 3ft 9in (1.14m).
Launch weight: 1,100lb (500kg).

Development and Service

Although Walleye is unpowered, it was once described by the US Navy as the most accurate and effective air-to-surface conventional weapon ever developed anywhere. The Naval Weapons Center at China Lake, California, was responsible for its development, the idea being to produce a 'smart bomb' that would find its own way unerringly to the target after the launch aircraft had turned for home. Martin Marietta received their first production contract in January 1966, their assembly lines being supplemented by others at Hughes Aircraft Company for a period, to meet the demands of the war in Vietnam. Since the Spring of 1971, Martin Marietta has been developing a Walleye 2 version, with a larger warhead.

Walleye is carried by a variety of aircraft, including the A-6, A-7, F-4, F-105 and F-111. Before launch, the pilot focuses the missile's TV camera on the target, with the aid of a TV monitor screen in the cockpit. Once the camera has been locked on to the target, mechanism inside the Walleye takes over, 'watching' a screen inside the bomb and correcting any deviation from the flight path to the target during the bomb's unpowered descent.

Photo Credits

Aviation Week and Space Technology 14, 106
Flight International 7, 127
Howard Levy 134
Indonesian Air Force 62 (lower)
Peter R. March 10
Ministry of Defence 21 (upper), 35 (lower)
Novosti 37 (lower), 48 (lower), 110 (lower), 115
Ronaldo S. Olive 75
Royal Australian Navy 58 (upper)
Royal Hellenic Air Force 84
Royal Netherlands Air Force 52 (lower)
Soviet Weekly 3, 36, 41, 46, 110 (upper), 116, 117, 119, 120 (upper), 153
Swedish Air Force 133
Tass 37 (upper), 38, 39, 42, 43, 45, 47, 48 (upper), 49, 108, 109, 111, 112, 113, 114 (lower), 118, 120 (lower), 128, 135, 137, 150
US Army 54, 68 (upper), 69
US Air Force 86, 94
US Navy 15 (lower), 70 (upper), 93, 126, 132, 148 (lower), 149, 151

Index

AAM-1 4
AAM-2 4
ACRA 4
ADM-20C Quail 94
AGM-12B/E Bullpup 23
AGM-12C/D Bullpup 24
AGM-22A, AS 11 10
AGM-28 Hound Dog 57
AGM-45A Shrike 132
AGM-53A Condor 28
AGM-65A Maverick 78
AGM-69A SRAM 142
AGM-78 Standard ARM 148
AIM-4 Falcon 33
AIM-4 Super Falcon 152
AIM-7 Sparrow 138
AIM-9 Sidewinder 1A 133
AIM-9 Sidewinder 1C 134
AIM-26 Nuclear Falcon 86
AIM-47A Falcon 33
AIM-54A Phoenix 90
AIR-2A Genie 44
AJ.168 Martel 76
AS-1 Kennel 62
AS-2 Kipper 63
AS-3 Kangaroo 61
AS-4 Kitchen 63
AS-5 Kelt 62
AS-11 10
AS-12 11
AS.20 12
AS.30 13
AS.30L 13
AS-37 Martel 76
ATM-1 type 64, KAM-3D 60
Advanced Terrier 5
Aegis 6
Agile 7
Albatros 7
Albatros 8
Alkali 8
Anab 9
Ash 14
Asroc 15
Atoll 16
BO 810, Cobra 2000 27
Bantam 17
Bloodhound 18
Blowpipe 20
Blue Steel 21
Bomarc 22
Bulldog 23

Bullpup A 23
Bullpup B 24
CF.299 Sea Dart 122
CGM-13 Mace 74
CIM-10B Bomarc 22
Cactus 29
Chaparral 25
Chinese ICBM 26
Chinese MRBM 26
Cobra 2000 27
Condor 28
Crotale 29
Dragon 30
Entac 31
Exocet 32
Extended Range Lance 73
FOBS Scarp 115
Falcon AIM-4A/C/D/H 33
Falcon AIM-47A 33
Falcon HM-55/HM-58 34
Firestreak 35
Frog-1 36
Frog-2, 3, 4, 5, 36
Frog-7 38
Gabriel 40
Gainful 41
Galosh 42
Ganef 43
Genie 44
Goa 45
Griffon 46
Guideline 47
Guild 49
HARM 50
HM-55 Falcon 34
HM-58 Falcon 34
HOBOS 53
HOT 56
Hardsite 49
Harpon 50
Harpoon 51
Hawk 52
Hellcat 53
Honest John 54
Hornet 55
Hound Dog 57
Ikara 58
Indigo 59
KAM-3D 60
KAM-9 61
Kangaroo 61
Kelt 62

Kennel 62	RIM-24 Tartar 156
Kipper 63	RIM-66A/67A Standard Missile 149
Kitchen 63	RUR-5A Asroc 15
Kormoran 64	Rapier 97
LGM-25C Titan 160	Redeye 102
LGM-30 Minuteman 80	Red Top 103
Lance 73	Roland 104
M47 Dragon 30	SA-2 Guideline 47
MD-660 78	SA-3 Goa 45
MGM-29A Sergeant 129	SA-4 Ganef 43
MGM-31A Pershing 88	SA-6 Gainful 41
MGM-32A Entac 31	SAM-D 106
MGM-51A Shillelagh 131	SA-N-1 Goa 45
MGM-71A TOW 160	SA-N-2 Guideline 47
MGR-1 Honest John 54	SA-N-3 45
MILAN 79	SCAD 113
MIM-3 Nike Ajax 84	SLAM 136
MIM-14 Nike Hercules 85	SRAAM 75 141
MIM-23A Hawk 52	SRAM 142
MIM-43A Redeye 102	SS-1b Scud-A 118
MIM-72A Chaparral 25	SS-1c Scud-B 120
MM-38 Exocet 32	SS-4 Sandal 109
MSBS 83	SS-5 Skean 135
MTM-3 Pershing 88	SS-8 Sasin 110
Mace 74	SS-9 Scarp 115
Magic 97	SS.10 144
Malafon 75	SS.11 145
Martel 76	SS-11 145
Masurca Mk 2 77	SS.12 146
Matador 74	SS-12 Scaleboard 113
Maverick 78	SS-13 Savage 111
Minuteman 80	SS-14 Scamp/Scapegoat 114
Mosquito 82	SSBS 147
Murène 84	SS-N-? 147
Nike Ajax 84	SS-N-1 Scrubber 117
Nike Hercules 85	SS-N-2 Styx 150
Nord 5203, SS 10 144	SS-N-3 Shaddock 130
Nuclear Falcon 86	SS-N-4 Sark 110
Otomat 87	SS-N-5 Serb 128
Penguin 88	SS-N-6 Sawfly 112
Pershing 88	Safeguard 104
Phoenix 90	Sagger 106
Pluton 91	Samlet 108
Polaris 92	Sandal 109
Poseidon 93	Sark 110
Quail 94	Sasin 110
R.511 95	Savage 111
R.530 96	Sawfly 112
T.550 Magic 97	Scaleboard 113
RB04 99	Scamp 114
RB05A 100	Scapegoat 114
RB07 Seacat 121	Scarp 115
RB08A 101	Scrag 116
RB27, RB28 Falcon 34	Scrooge 117
RB68 Bloodhound Mk 2 18	Scrubber 117
RB69 Redeye 102	Scud-A 118
RIM-2 Advanced Terrier 5	Scud-B 120
RIM-7H Sea Sparrow 126	Seacat 121
RIM-8G-AAW Talos 155	Sea Dart 122
RIM-8H-ARM Talos 155	Sea Indigo 123

Sea Killer Mk 1 123
Sea Killer Mk 2 124
Seaslug 125
Sea Sparrow 126
Seawolf 127
Serb 128
Sergeant 129
Shaddock 130
Shillelagh 131
Shrike 132
Sidewinder 1A 133
Sidewinder 1C 134
Skean 135
Snapper 137
Sparrow 138
Spartan 139
Sprint 140
Standard ARM 148
Standard Missile 149
Styx 150
Subroc 151
Super 530 152
Super Falcon 152
Swatter 153

Swingfire 154
TAN-SSM, KAM-9 61
TM-61 Matador 74
TOW 160
Talos 155
Tartar 156
Terne Mk 8 157
Terrier 5
Thunderbird 159
Tigercat 159
Titan 11 160
UGM-27 Polaris 92
UGM-73 Poseidon 93
ULMS 162
UUM-44A Subroc 151
V750VK Guideline 47
Vigilant 162
WS-140A SRAM 142
Walleye 163
XLIM-49A Spartan 139
XMGM-52B Lance 73
XRL Lance/Extended Range Lance 73
ZAGM-64A Hornet 55
ZAGM-84A Harpoon 51